Introduction to Astronomy and Photography

Dr. John A. Allocca
Copyright 2017. Updated: 9/9/23

Published by
Allocca Biotechnology, LLC
www.allocca.com
www.DrJohnPhotography.com

ISBN 978-1547107346

Dr. Allocca is the innovator of:

- Neurochemical Reprogramming for for Migraine, Depression, and More

- Brainicity™ E.L.F. Brain Biofield Enhancement

- Orthomolecular Assessment and Wellness Plan

- Photography - A Path To Healing (Books, Classes, Workshops, Field Trips)

Table of Contents

Part 2 - Photography Basics ---------71

Part 1 - Astronomy Basics

Introduction

Astronomy has been a wonderful and exciting aspect of human civilization since the dawn of time. Start a new adventure in your life with astronomy.

Telescopes and binoculars provide you with a close view of the stars. An inexpensive telescope, spotting scope, or pair of binoculars can provide a good view of the moon, planets, and the Andromeda Galaxy (M31).

When you look up at the stars with the naked eye, you may see six stars in a given area. When you look at the same area with a telescope or binocular, you may see ten times or more than that.

There is backyard astronomy and there are star gazing parties and events. Check your local area.

There is a huge selection of optics on the market. Some will be covered herein.

Learning Astronomy

To begin, watch the 46 part, 10-15 minute segment, Crash Course Astronomy presentation by Dr. Phil Plait of PBS at youtube: https://www.youtube.com/watch?v=0rHUDWjR5gg

Subscribe to Astronomy Magazine and/or Sky and Telescope Magazine. There are many books, magazines, and courses available in astronomy. Check your local library or Amazon.com.

Binoculars

Binoculars are probably the easiest way to learn astronomy. Simply aim at the stars. The image seen through binoculars is upright. Therefore, easy to follow. Binoculars can also be simple and inexpensive or highly advanced and expensive.

Try this experiment. Look up and around the room you are in. Close one eye. Then, open it. Try this several times. The depth perception with both eyes is much greater than the depth perception with one eye. It is the difference between 2D and 3D vision. This is the difference between viewing through binoculars and telescopes.

Binoculars are specified as power and aperture. The aperture is the width of the optical lens. For example, a 7 x 50 binocular would have a power of 7 times that of the normal eye with an aperture of 50 mm. Seven power is usually the limit of power that can be held by hand without any support. Higher power will require a tripod, monopod, or other device. The larger the aperture, the more you can see, and the heavier it will be. A binocular of 10 x 50 will reveal much more than a 7 x 50. Some kind of bracing with ones elbows may work to replace a tripod.

Generally, fixed eyepiece binoculars have lower power capabilities than telescopes. Most binocular powers range from 7x to 25x. Common telescopes often range to 300x.

Binoculars are good for observing the moon, planets, nebula, and large galaxies like M31 (Andromeda). Telescopes can be used for deep space observation (DSO) at higher power. Higher power will require a tripod and tracking capabilities to keep up with the earth's rotation, which will be discussed later.

The higher the power, the more difficult it will be to hold steady. Several manufacturers offer image stabilized binoculars. They use electronic circuitry to stabilize the image. They are generally more expensive than standard binoculars.

Eye relief is the distance between the optics and your eye. If you wear glasses, you want a long eye relief of at least 15 mm. You may need to fold the rubber eye cups down if you wear glasses. Some binoculars have helical or flip up eye cups that raise and lower by twisting the eyecups.

The distance between the center of your eyes varies from person to person. This is known as the Interpupillary Distance (IPD). When looking through a pair of binoculars the ocular lenses must line up perfectly with your pupils. If not, you will see a dark halo forming around the image and only see a small part of the image you are viewing. Most binoculars use a hinge system between the two barrels allowing you to adjust the barrels. With some binoculars, only the eyecup is moved.

Some tripods that can be used are the Celestron Trail Seeker Tripod ($100) and the Celestron Regal Premium Tripod ($150). The Regal tripod is heavier and sturdier, which may be required for heavier binoculars. 10x50 binoculars are popular, however they will most likely require a tripod or monopod. The most common binoculars used for astronomy are 15x70, 20x60, 20x80, and 25x100. These large binoculars will definitely require a tripod or monopod. There are also parallel arm tripod assemblies made for use with binoculars such as the Orion Paragon-Plus Binocular Mount and Tripod that can be lifted or lowered to any position without changing the angle of the binocular ($250). Some people lay in a zero gravity lounge chair while viewing the sky with binoculars. The Celestron SkyMaster Pro binoculars DO NOT accept 1.25" filters as some websites state.

To advance further, there are the Orion BT-100 binoculars with 90 degree angled 1.25" interchangeable eyepieces ($1,200). There is also the Vixen Optics 25x81 BT81S-A Astronomical Binocular with 45 degree angled 1.25" interchangeable eyepieces ($2,000). Interchangeable eyepieces will allow for higher power and accept 1.25" filters. Both require the use of a tripod.

A good binocular to start astronomical observations with would be a 7x50 binocular. Generally, the more expensive the binocular, the better is the quality. Generally, binoculars that cost less than $100 are only fair in quality.

Some binoculars, weights, and current prices:
Canon 15x50 IS Image Stabilized, 21.2 oz. (2.7 pounds), $1,000
Celestron Outland X 8 x 42, 22 oz. (1.38 pounds), $60
Celestron Cometron 7 x 50, 27.3 oz. (1.70 pounds), $25

Nikon Aculon 7 x 50, 31.9 oz. (1.99 pounds), $100
Nikon Aculon 16 x 50, 32.6 oz. (2.04 pounds), $120
Nikon PROSTAFF 5 10x50 (1.8 pounds), $200
Nikon Model MONARCH 3 10x42 ATB (1.5 pounds), $250
Pentax 10 x 50 SP WP, 37.4 oz. (2.19 pounds), $250
Pentax 20 x 60 SP WP, 49.4 oz. (3.09 pounds), $250
Celestron Sky master 20 x 80, 94.24 oz. (5.86 pounds), $115
Celestron Skymaster Pro 20 x 80, 86.4 oz. (5.4 pounds), $250
Vixen 20 x 80, 84.8 oz. (5.3 pounds), $700

The Unimount and Tripod in the reclining position with a zero gravity or other chair is also a great method for viewing using higher powered binoculars.

Never look at the sun with binoculars.

Binocular Collimation

Basically, binoculars are two telescopes side by side. Each telescope presents an image to each eye. These two telescopes must be perfectly aligned (collimated) in parallel. If they are not, the images will be misaligned in the eyepieces at which point the brain will try to adjust them to a single image. The result may be eye strain and headaches. Binoculars can easily become misaligned if they are dropped or mishandled.

If binoculars are misaligned, highly trained people and equipment are required to properly align them.

To test binoculars for alignment do the following:

1. Point the binoculars at a straight horizontal line in the distance. It could be a roof or power line.

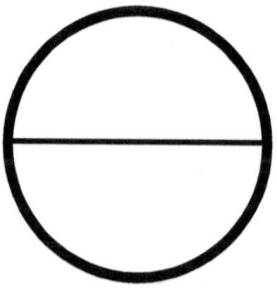

Image with eyes close to the binoculars

2. Slowly bring the binoculars 2-3 inches away from your eyes. The image will separate into two circles. If the binoculars are aligned, there will be a straight line through the two circles. If they are not aligned, there will be two separate lines as can be seen below.

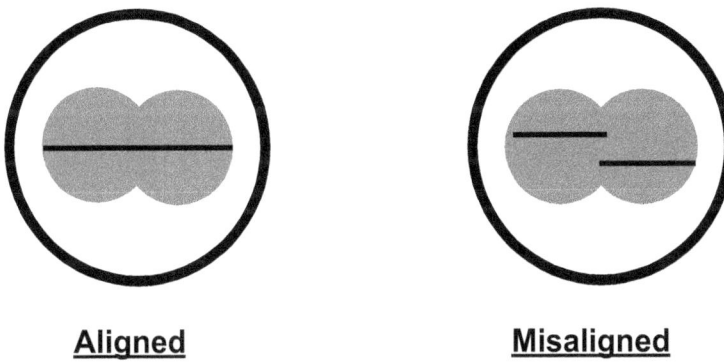

<u>Aligned</u>   **<u>Misaligned</u>**

Telescopes and Spotting Scopes

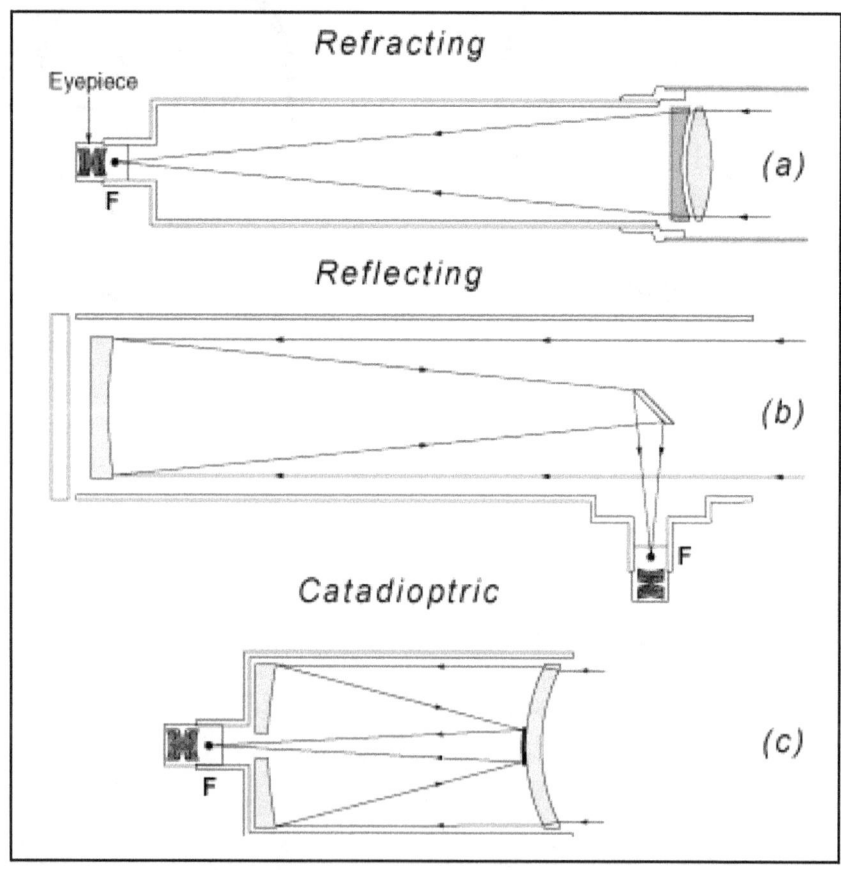

Generally, the difference between a telescope and a spotting scope is that the image in a spotting scope is upright. The image in a telescope is upside down. An upright image will be easier for a beginner to relate to and navigate the night sky. Telescopes come primarily in three varieties: refractor, reflector, and catadioptric.

The refracting telescope is the simplest type of telescope that uses lenses to form an image. It is the oldest telescope sometimes also called Galilean telescope. The refractor telescope requires little or no maintenance. If the lenses are good, refractor telescopes give a good high-contrast image. They are especially desirable for viewing the moon and planets. Refractor telescopes generally have small openings, usually between 60 mm and 120 mm. For many astronomical purposes, this aperture is very small. Little bright objects like galaxies and nebulae will appear as weak blurs when you can actually detect them. Poor quality lenses will produce chromatic aberrations (color fringing or purple fringing).

The reflector is also known as the Newtonian telescope. These use a large, heavy concave mirror instead of lenses to collect and focus light. The light from distant objects that enter the telescope tube are reflected in the concave mirror to a flat diagonal mirror. The diagonal mirror then reflects the light through an opening in one side of the telescope eyepiece lens. Compared to other telescopes of the same opening, a reflector telescope is the least expensive telescope. An added advantage lies in the fact that the reflectors work with mirrors instead of lenses, therefore, they do not produce chromatic aberrations. They are bulkier than refractors and their management is less intuitive, since the eye is located near the top of the telescope.

The catadioptric telescope uses a lens that is placed at the end of the telescope. Light passes through the first and then reaches the primary mirror at bottom of tube which, reflects the light to a secondary mirror. It is a blend of the refractor and reflector telescopes. They are short and heavy but easy to transport because of their short length tubes. Their optical quality is good but it fails to outperform a good refractor telescope. They are more expensive than the reflector telescopes of the same aperture. The most common catadioptric telescopes are Schmidt-Cassegrain and Maksutov-Cassegrain. The Schmidt-Cassegrain, light enters through a thin aspheric Schmidt correcting lens, then strikes the spherical primary mirror and is reflected back up the tube to a small secondary mirror. The mirror then reflects the light out an opening in the rear of the instrument where the image is formed at the eyepiece. The Maksutov-Cassegrain telescope uses a thick meniscus-correcting lens with a strong curvature and a secondary mirror that is usually an aluminized spot on the corrector. The Maksutov secondary mirror is typically smaller than the Schmidt's giving it slightly better resolution for planetary observing.

You can start with a spotting scope and a photographic tripod. The larger the aperture, the more you can see, and the heavier it will be. Spotting scopes are easy to use and don't require alignment at low power. They are also very light weight. Some options may be the Celestron C70, 70 mm Maksutov-Cassegrain ($120), Celestron C90, 90 mm Maksutov-Cassegrain ($230), or the Celestron C5, 5 inch Maksutov-Cassegrain ($600). The later three spotting scopes accept 1.25" eyepieces and filters. A decent photographic tripod is the Celestron Astro Master alt-azimuth mount, which has a 1/4-20 mount ($80) or the Celestron heavy-duty alt-azimuth mount tripod, which has a 1/4-20 mount ($100). The alt-azimuth mount tripod will allow you to manually track the stars, which may be needed at higher power. An inexpensive equatorial mount tripod is the Orion EQ-1 ($124). Heavier and sturdier equatorial mounts get more expensive as you increase in size. Tripods and mounts will be explained later.

Another lightweight telescope to consider is an Orion 80 mm Short Tube refractor telescope ($200) with an equatorial mount tripod, such as the Celestron heavy-duty alt-azimuth mount ($100). This setup requires manual tracking. Any of the Celestron beginner telescopes are also recommended. You may want to consider a computerized telescope.

The earth rotates and so does your view of the sky. Telescopes are usually aligned to polaris (north). Then, the motor mechanisms will rotate the telescope to follow the path of the stars. Or, one can turn the telescope manually with knobs to follow the path.

A more advanced telescope with a computerized (GOTO) function, which will automatically point to a specific star, would be the Celestron Evolution series. The telescope is advanced, but can be used by beginners. They are available in 6", 8", and 9.25" models and cost $1,200 to $2,100. The larger the telescope, the heavier it will be. The Celestron Evolution series telescopes work with the SkyPortal iPhone and Android apps.

Never look at the sun with a telescope or spotting scope without a solar filter.

Magnification

The magnification or how much an object is enlarged by a telescope is calculated by dividing the focal length of the telescope by the eyepiece. For example, if the telescope focal length is 400 mm and you are using a 25 mm eyepiece, the magnification is 400/25 = 16x. The useful magnification for a telescope is two times the aperture in millimeters. So, for a 80 mm refractor, the maximum useful power is 160x. Higher magnification can cause the image to appear dim and fuzzy. The magnification needed for viewing planets is usually 100x or more.

Never look at the sun with a telescope or spotting scope without a solar filter.

Field of View

As magnification increases, the field of view decreases. The field of view depends upon the telescope and the eyepieces used. An apparent field of view varies with eyepieces. A 25 mm eyepiece may have a field of view of 50 degrees. Another 25 mm eyepiece may have a field of view of 70 degrees. The later eyepiece will show a larger area with the same magnification.

To determine the actual field of view that each eyepiece produces for a specific telescope, divide the apparent field of view of the eyepiece by the magnification the eyepiece produces in a telescope.

Example 1, a telescope with a focal length of 400 mm and an eyepiece of 25 mm with a field of view of 50 degrees will produce a magnification of 16x. Therefore, divide 50 degrees by 16x = 3.13 degrees field of view.

Example 2, a telescope with a focal length of 400 mm and an eyepiece of 12.5 mm with a field of view of 50 degrees will produce a magnification of 32x. Therefore, divide 50 degrees by 32x = 1.56 degrees field of view.

Example 3, a telescope with a focal length of 400 mm and an eyepiece of 12.5 mm with a field of view of 70 degrees will produce a magnification of 32x. Therefore, divide 70 degrees by 32x = 2.19 degrees field of view.

Never look at the sun with a telescope or spotting scope without a solar filter.

Sky Angles can be estimated by holding ones hand out at arms length as follows:

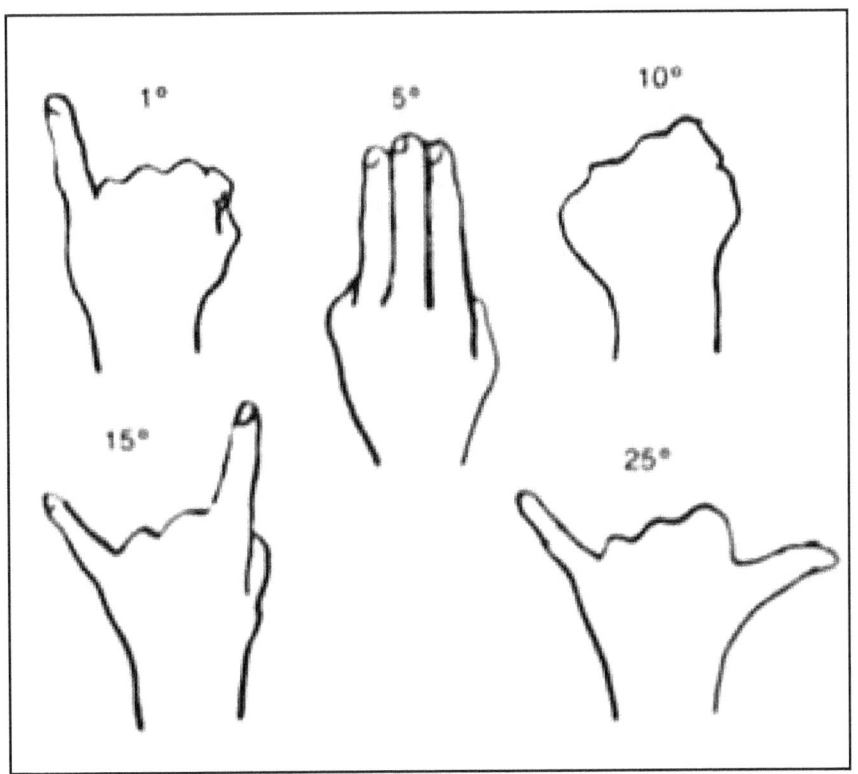

Light Gathering

The area of a circle = Pi x Radius squared. The greater the area, the move light gathered. An 8" optical tube will gather almost twice as much light as an 6" optical tube.

60 mm (2.36"): Area = 2,826 mm
70 mm (2.76"): Area = 3,846.5 mm
80 mm (3.15"): Area = 5,024 mm
102 mm (4.02"): Area = 8,167.1 mm
127 mm (5.00"): Area = 12,661.3 mm
153 mm (6.02"): Area = 18,376.1 mm
204 (mm 8.03"): Area = 32,668.6 mm

Tripods and Mounts

The photographic tripod or alt-azimuth mount is the simplest mount. The alt-azimuth mount has two axes of rotation, a horizontal axis and a vertical axis. To point the telescope at an object, rotate it along the horizon (azimuth axis) to the object's horizontal position, and then tilt the telescope, along the altitude axis, to the object's vertical position. It does not align and track stars. This type of mount can be used for binoculars, spotting scopes and low power telescopes.

The equatorial mount also has two perpendicular axes of rotation: right ascension and declination. However, instead of being oriented up and down, it is tilted at the same angle as the Earth's axis of rotation. This type of mount can track stars manually or automatically when connected to a clock mechanism.

Various tripods and mounts are available for binoculars too. Orion makes parallel mounts that can be lifted or lowered to any position without changing the angle of the binocular. Use a reclining lounge chair with a parallel mount for an evening of relaxation and enjoyment of the night sky.

Filters

Eyepiece filters are used primarily in lunar and planetary observing. They reduce glare and light scattering, increase contrast, and increase resolution. Filters screw into the bottom of eyepieces.

Moon Filters cut down glare and bring out surface detail with better contrast. They usually transmit 13% of light.

Planetary filters block out certain colors in the visible spectrum of light. A red filter will block out all but the red wavelength of light. If you look at an object that is primarily red while using a red filter, the object will appear very bright. Areas which are not red will appear more clearly because they contrast with the wavelength of light which is being passed by the filter.

Light Yellow (#8 Light Yellow) helps to increase the detail in the maria on Mars, enhance detail in the belts on Jupiter, increase resolution of detail in large telescope when viewing Neptune and Uranus, and enhance detail on the moon in smaller scopes.

Yellow-Green (#11 Yellow Green) helps to bring out dark surface detail on Jupiter and Saturn, darkens the maria on Mars, and improves visual detail when viewing Neptune and Uranus through large telescopes.

Yellow (#12 Yellow) helps greatly in viewing Mars by bringing out the polar ice caps, enhancing blue clouds in the atmosphere, increasing contrast, and brightening desert regions. Yellow also enhances red and orange features on Jupiter and Saturn and darkens the blue festoons near Jupiter's equator.

Orange (#21 Orange) helps increase contrast between light and dark areas, penetrates clouds, and assists in detecting dust storms on Mars. Orange also helps to bring out the Great Red Spot and sharpen contrast on Jupiter.

Light Red (#23A Light Red) helps to make Mercury and Venus stand out from the blue sky when viewed during the day. Used in large telescopes, light red sharpens boundaries and increases contrast on Mars, sharpens belt contrast on Jupiter, and brings out surface detail on Saturn.

Red (#25A Red) provides maximum contrast of surface features and enhances surface detail, polar ice caps, and dust clouds on Mars. Red also reduces light glare when looking at Venus. In large telescopes, a red filter sharply defines differences between clouds and surface features on Jupiter and adds definition to polar caps and maria on Mars.

Dark Blue (#38A Dark Blue) provides detail in atmospheric clouds, brings out surface phenomena, and darkens red areas when viewing Mars. Dark blue also increases contrast on Venus, Saturn, and Jupiter in large scopes.

Violet (#47 Violet) is recommended only for use on large telescopes. A violet filter enhances lunar detail, provides contrast in Saturn's rings, darkens Jupiter's belts reduces glare on Venus, and brings out the polar ice caps on Mars.

Light Green (#56 Light Green) enhances frost patches, surface fogs, and polar projections on Mars, the ring system on Saturn, belts on Jupiter and works as a great general-purpose filter when viewing the Moon.

Dark Green (#58 Green) increases contrast on lighter parts of Jupiter's surface, Venutian atmospheric features, and polar ice caps on Mars. Dark green will also help bring out the cloud belts and Polar Regions of Saturn.

Blue provides detail in atmospheric clouds on Mars, increases contrast on the moon, brings out detail in belts and polar features on Saturn, enhances contrast on Jupiter's bright areas and cloud boundaries. A blue filter is also useful in helping to split the binary star Antares when at maximum separation.

Light Blue functions much the same as #80A Blue while maintaining overall image brightness. Light blue will also help to increase structure detail when looking at galaxies.

Experiment with color filters. Set of filters to stat with is the Celestron Eyepiece Filter Set (1.25") ($40).

Polar Alignment with an Equatorial Mount

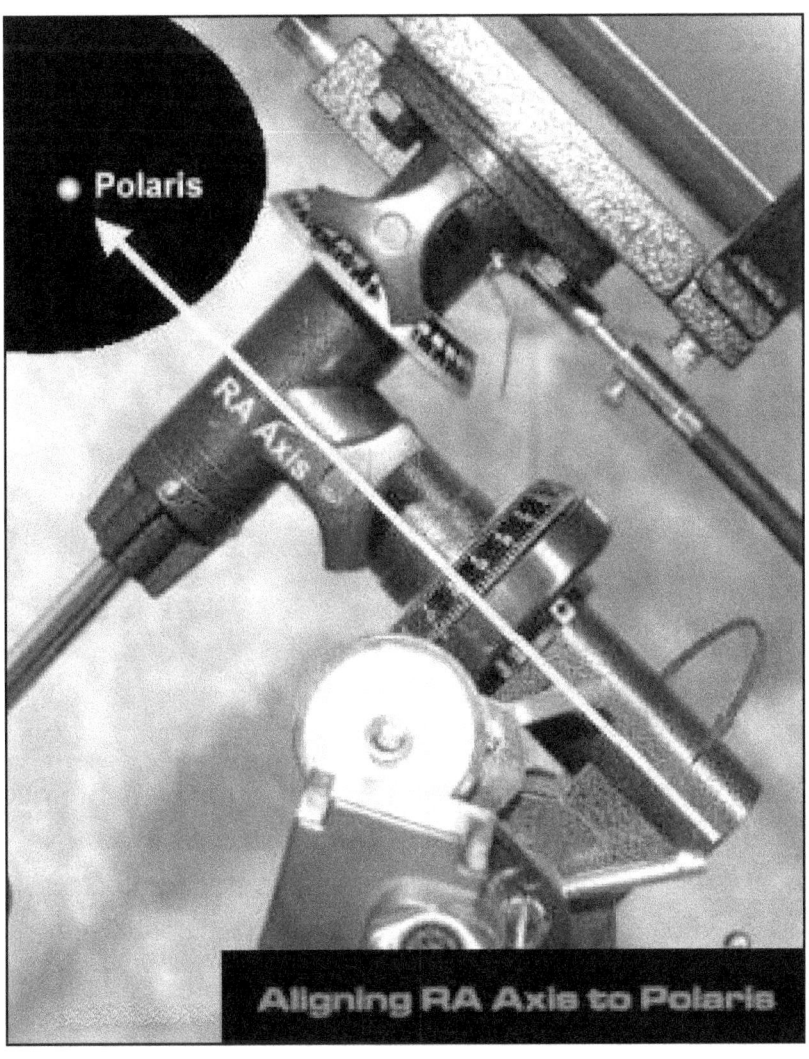

Aligning RA Axis to Polaris

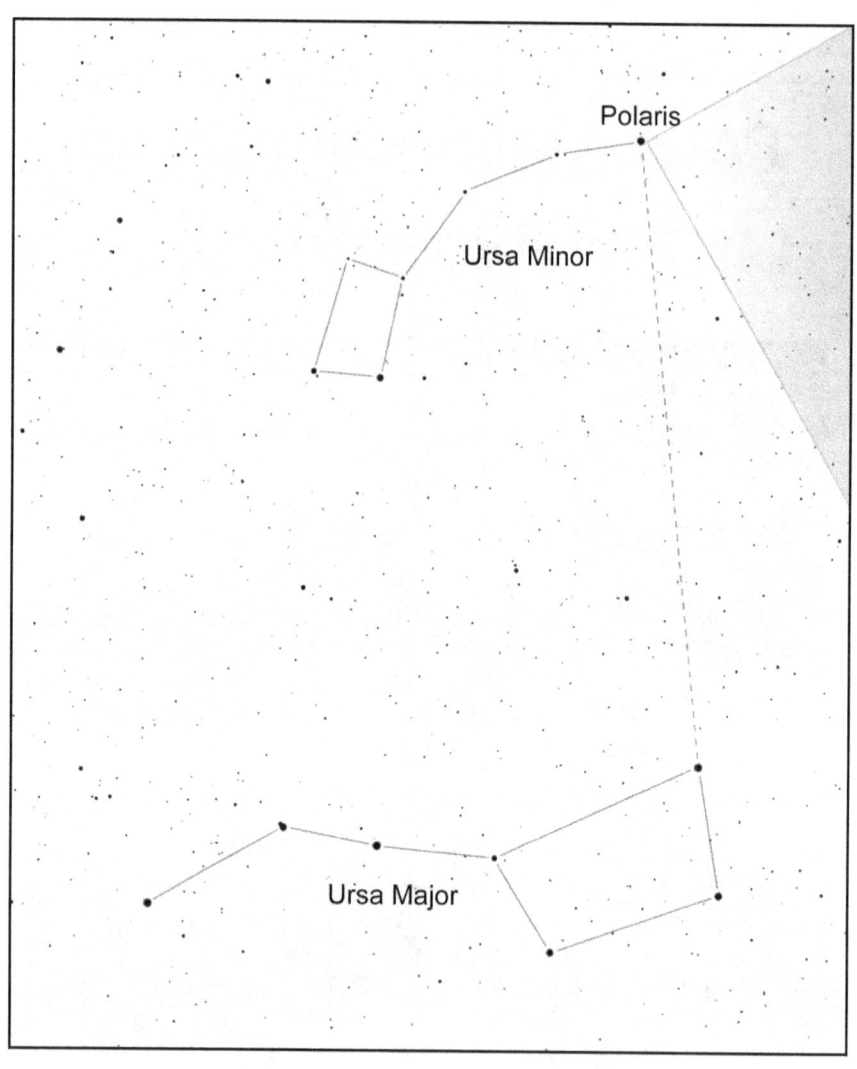

As the earth rotates the stars appear to cartwheel through the sky above. By aligning the telescope to a fixed point in the sky which isn't moving allows you to track objects using only the Right Ascension (RA) control. The Right Ascension movement compensates for the earths movement and allows the telescope to 'track' an object. The part of the sky which doesn't move is of the North Celestial Pole for the

Northern hemisphere. In the Northern Hemisphere Polaris is very close to the North Celestial Pole and provides a adequate position for observing. The basic aim of Polar Alignment is to align the telescope mounts Right Ascension (RA) axis to Polaris. The simplest method of polar alignment is simply to aim the RA axis at Polaris.

For general visual observation, an approximate polar alignment is sufficient:

1. Level the equatorial mount by adjusting the length of the three tripod legs.

2. Loosen the latitude lock t-bolt. Turn the latitude adjustment t-bolt and tilt the mount until the pointer on the latitude scale is set at the latitude of your observing site. If you don't know your latitude, consult a geographical atlas to find it. For example, Long Island, NY latitude is 41° North, set the pointer to +41. Then retighten the latitude lock t-bolt. The latitude setting should not have to be adjusted again unless you move to a different viewing location some distance away.

3. Loosen the Dec. lock thumb screw and rotate the telescope optical tube until it is parallel with the R.A. axis. The pointer on the Dec. setting circle should read 90°. Retighten the Dec. lock thumb screw.

4. Loosen the azimuth adjustment knob and rotate the entire equatorial mount left-to-right so the telescope tube (and R.A. axis) points roughly at Polaris. If you cannot see Polaris directly from your observing site, use a compass and rotate

the equatorial mount so the telescope points North. Retighten the azimuth adjustment knob.

The equatorial mount is now approximately polar-aligned for casual observing. More precise polar alignment is required for astrophotography. Several methods exist and are described in many amateur astronomy reference books and astronomy magazines.

From this point on in your observing session, you should not make any further adjustments to the azimuth or the latitude of the mount, nor should you move the tripod. Doing so will undo the polar alignment. The telescope should be moved only about its R.A. and Dec. axes.

Star Charts for the Northern Hemisphere

January - February

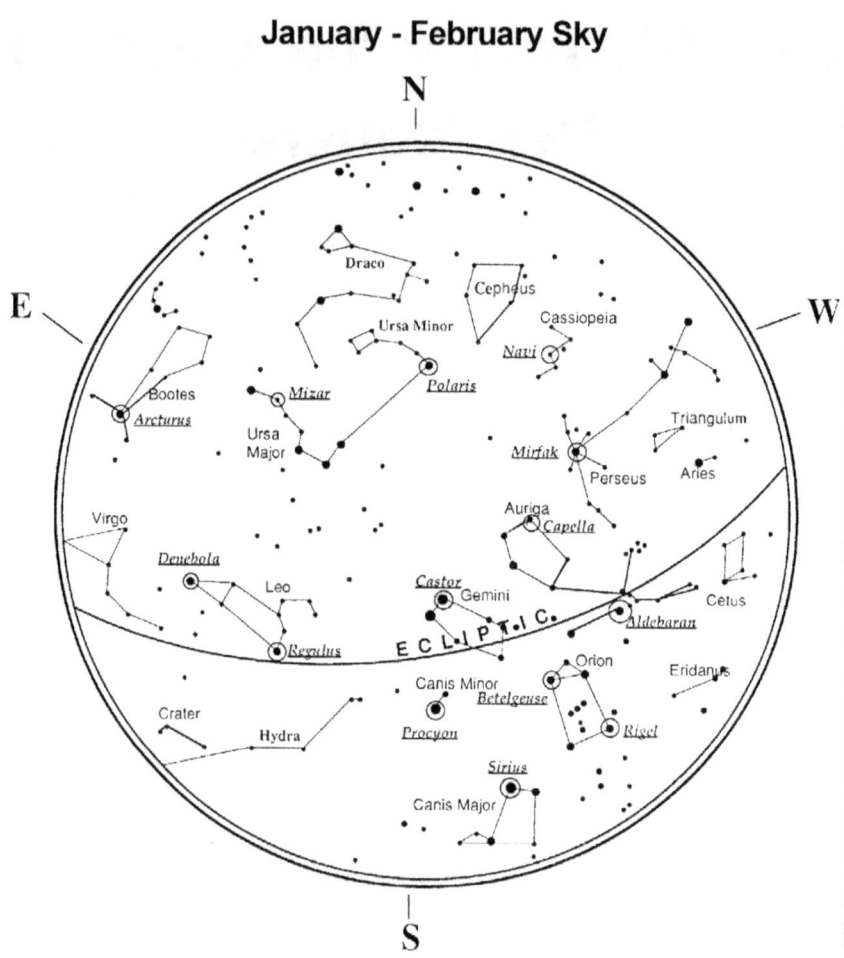

January - February Sky

March - April

March - April Sky

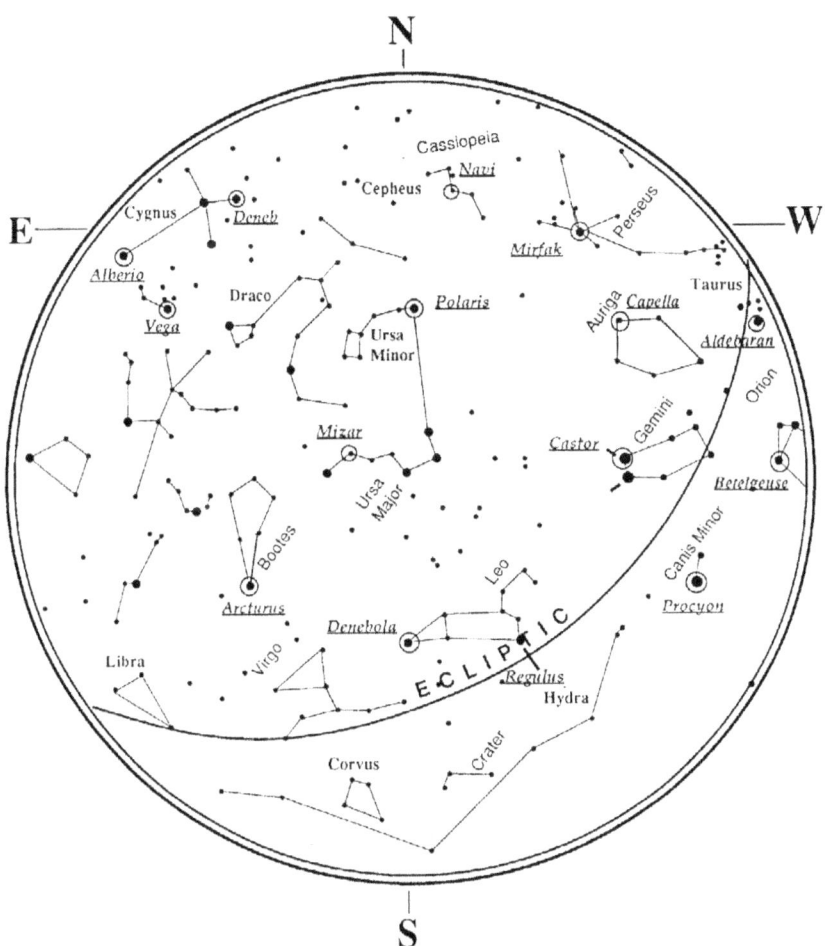

May - June

May - June Sky

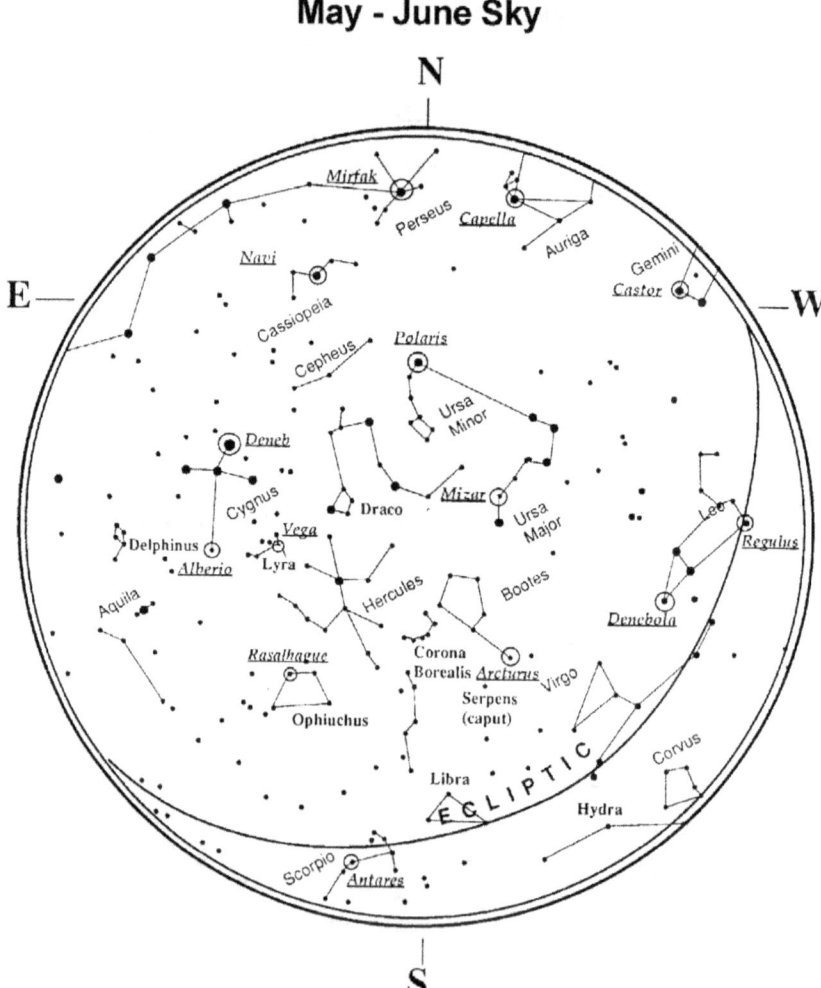

July - August

July - August Sky

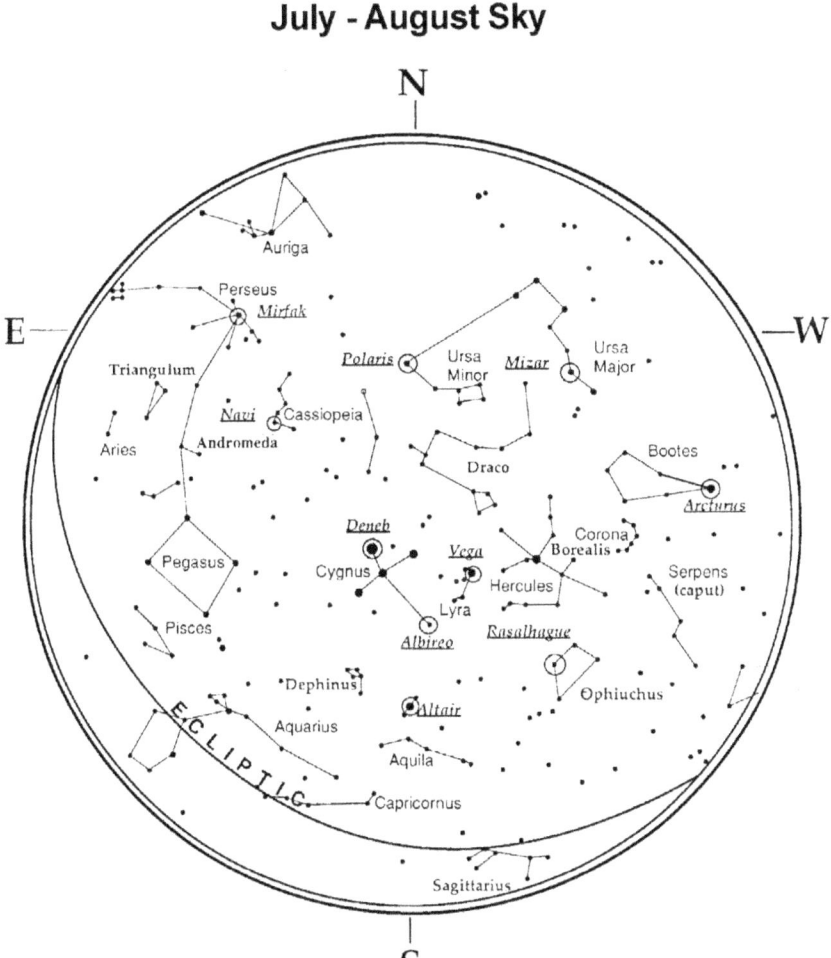

September - October

September - October Sky

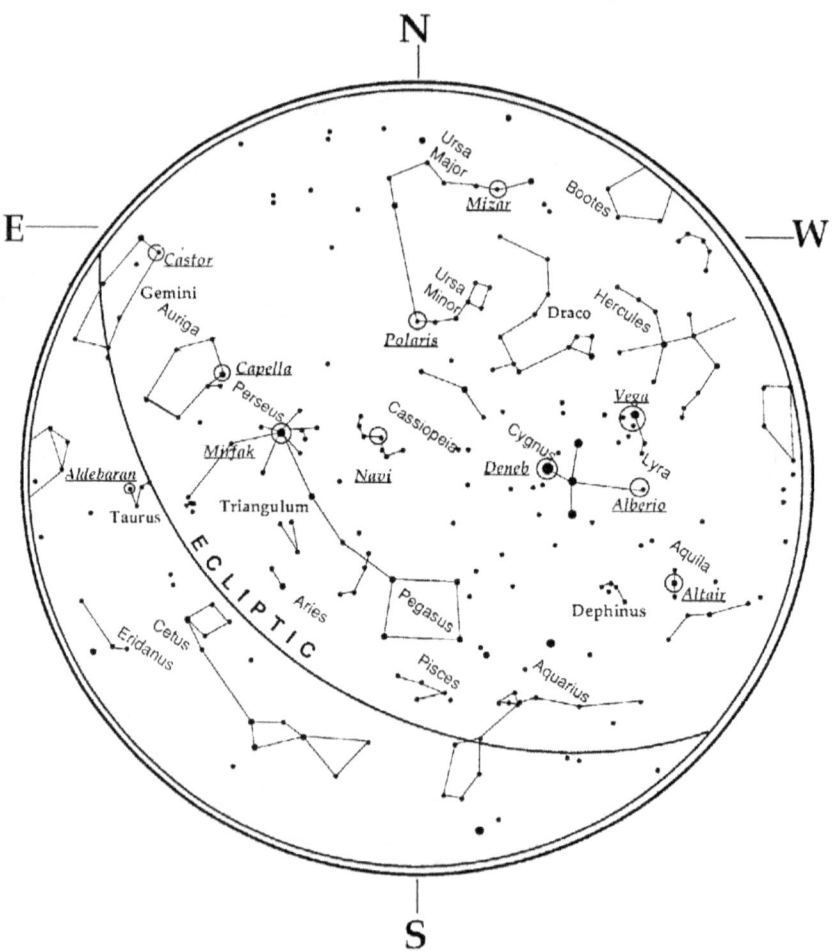

November - December

November - December Sky

N

Bootes
Hercules
Draco
Vega
Lyra
Ursa Major
Mizar
Alberio
E
W
Ursa Minor
Cygnus
Deneb
Polaris
Cassiopeia
Navi
Gemini
Perseus
Castor
Auriga
Mirfak
Pegasus
Capella
ECLIPTIC
Triangulum
Pisces
Procyon
Aries
Canis Minor
Betelgeuse
Aldebaran
Orion
Eridanus
Cetus
Rigel

S

41

Constellations

There are 88 recognized constellations within the northern and southern skies. The ancient significance and prediction of events of the constellations of the zodiac are not important to astronomers. The significance of the zodiac is that the ecliptic (the narrow path on the sky that the Sun, Moon, and planets appear to follow) runs directly through these constellations.

The 12 constellations of the mythological Zodiac in calendrical order are:

Ares, March 21 - April 20
Taurus, April 21 - May 21
Gemini, May 22 - June 21
Cancer, June 22 - July 22
Leo, July 23 - August 22
Virgo, August 23 - September 23
Libra, September 24 - October 23
Scorpio, October 24 - November 22
Sagittarius, November 23 - December 21
Capricorn, December 22 - January 20
Aquarius, January 21 - February 19
Pisces, February 20 - March 20

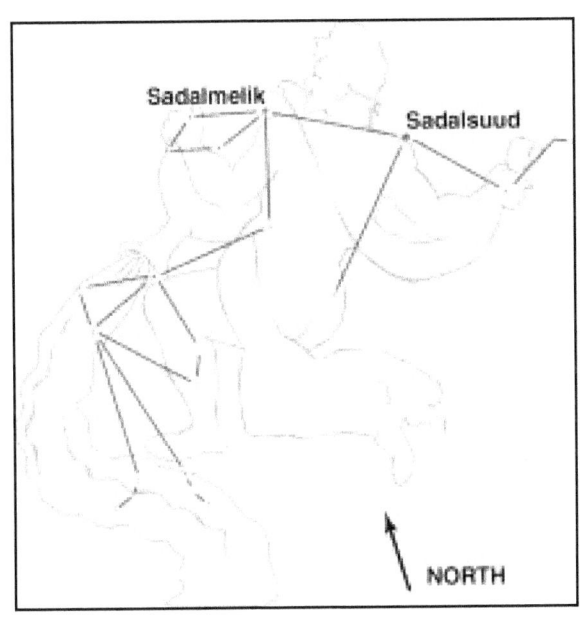

Aquarius

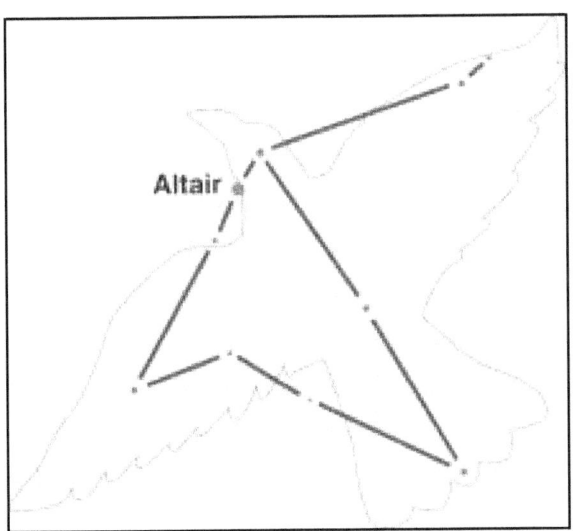

Aquila

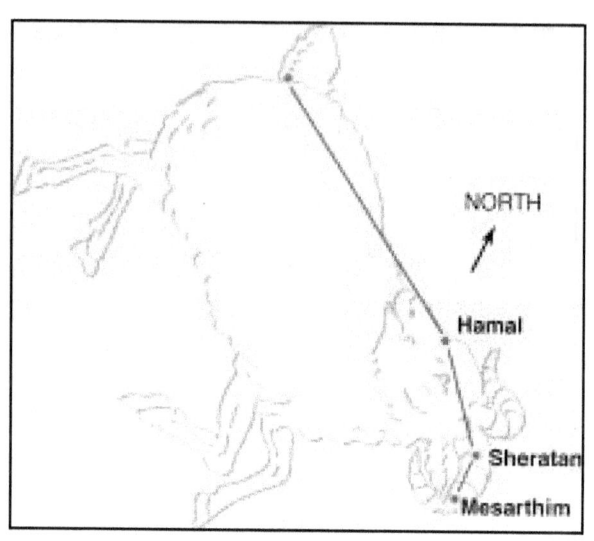

Aries

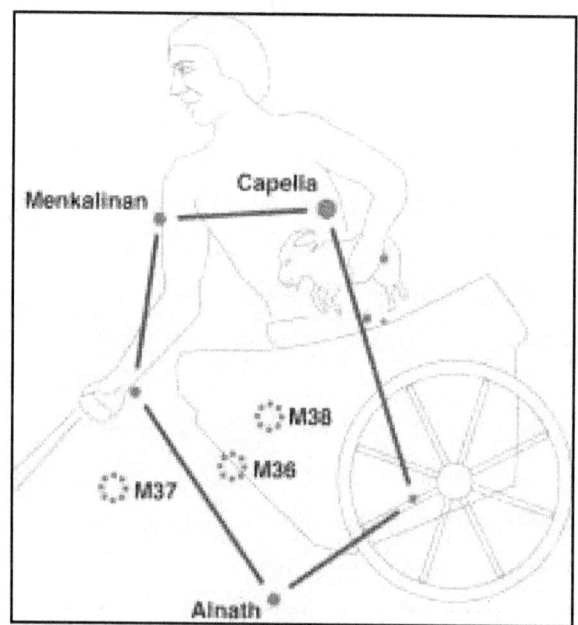

Auriga

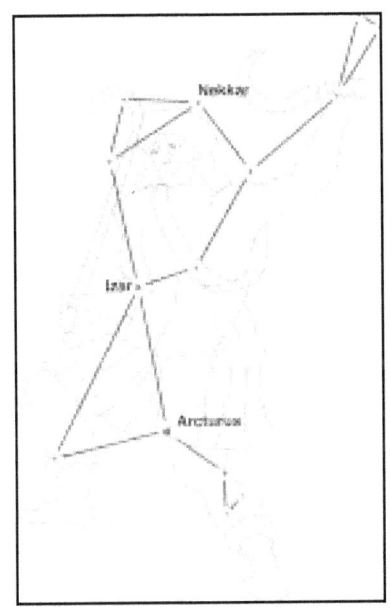

Bootes

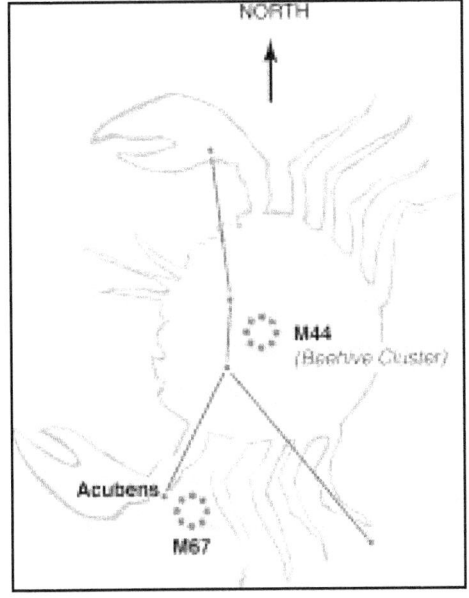

Cancer

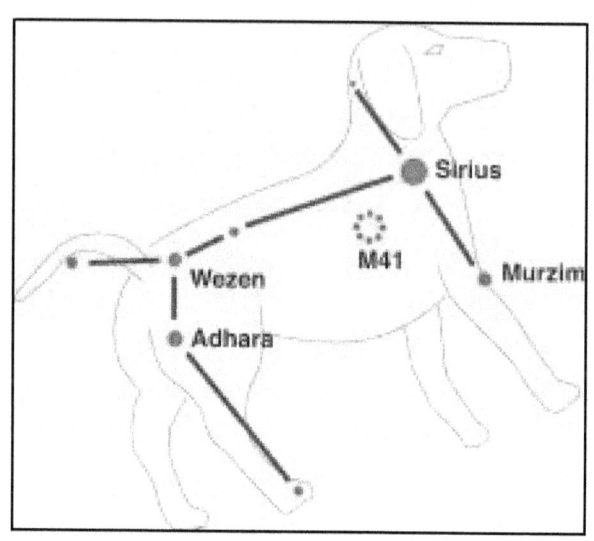

Canis Major

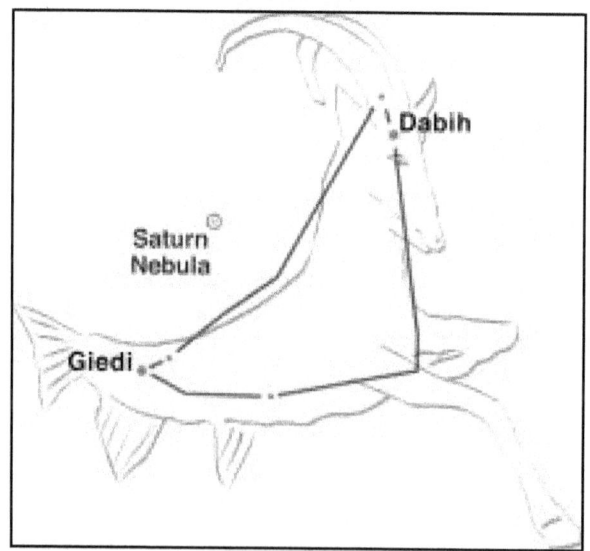

Capricornus

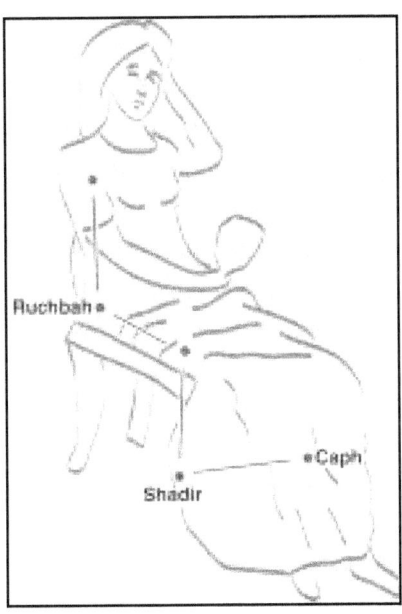

Cassiopeia

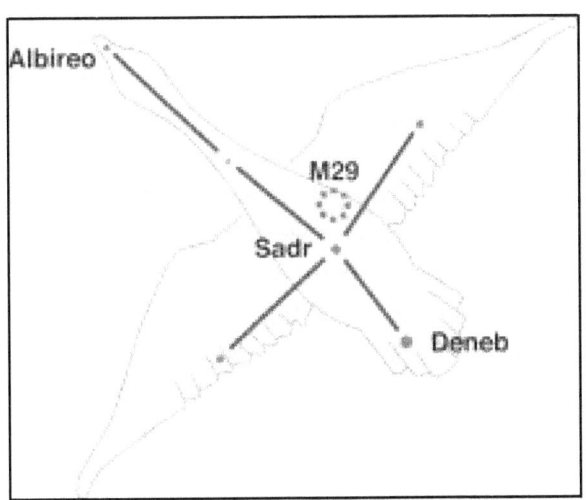

Cygnus

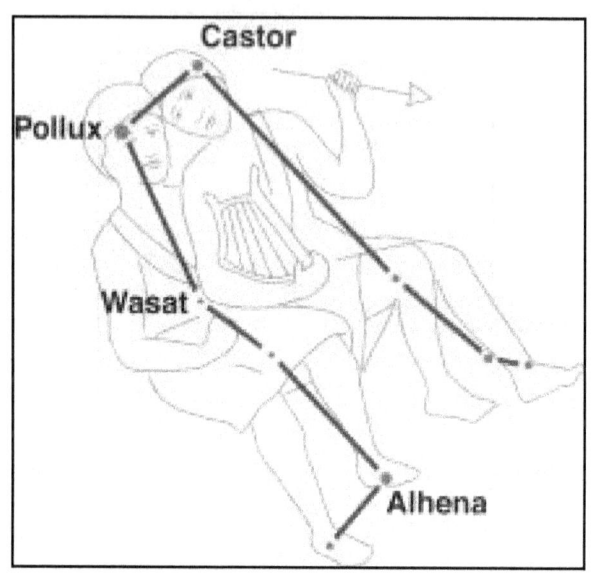

Gemini

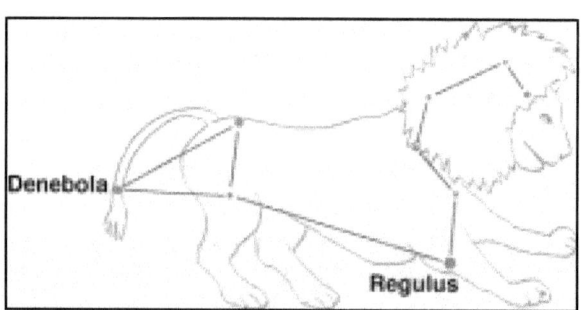

Leo

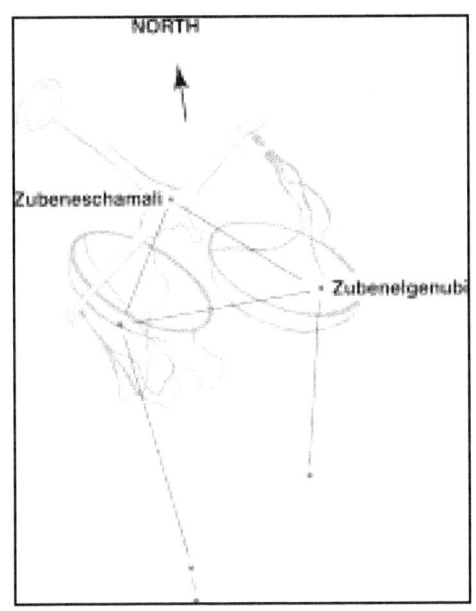

Libra

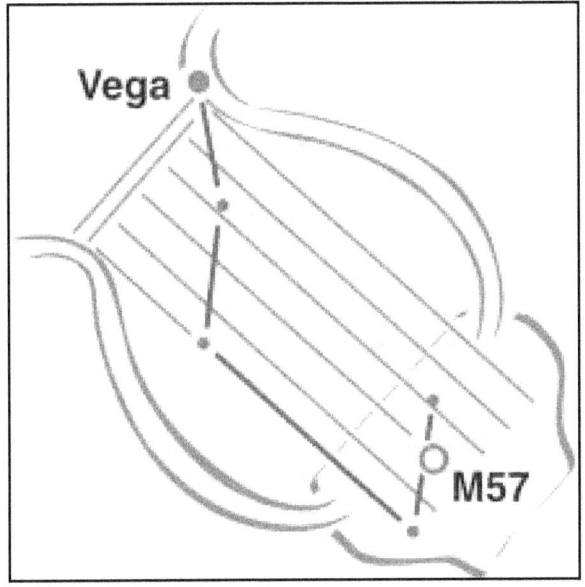

Lyra

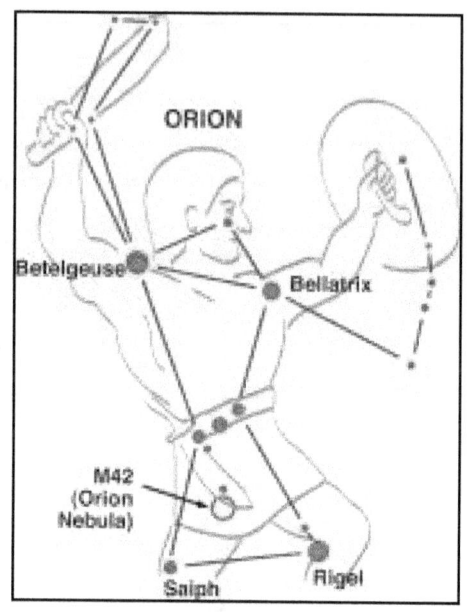

Orion

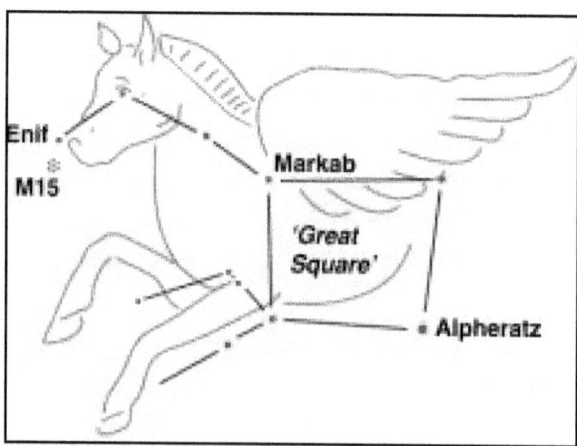

Pegasus

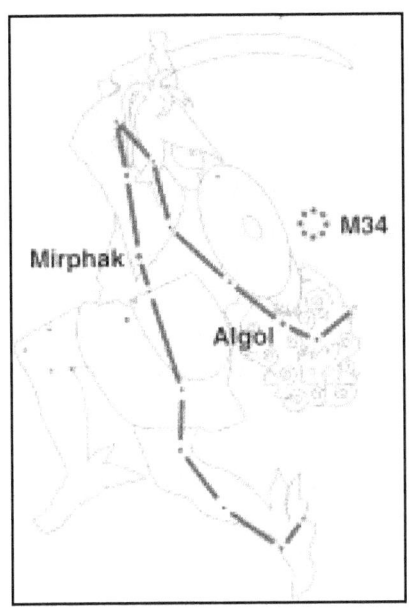

Perseus

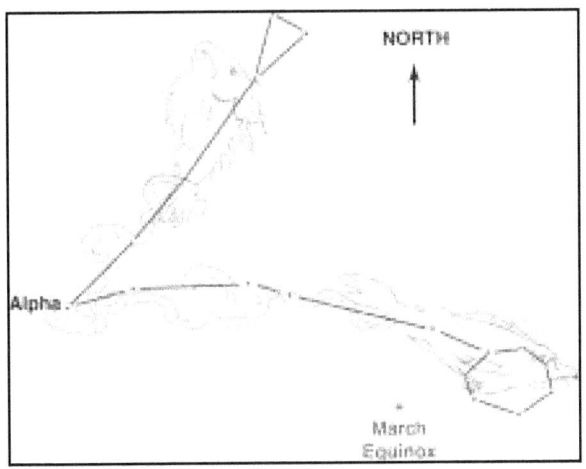

Pisces

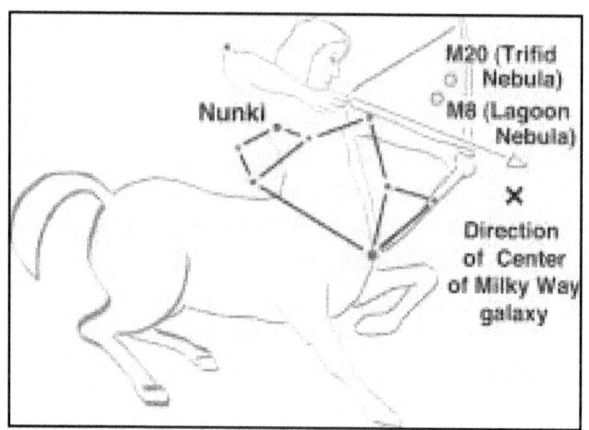

Sagittarius

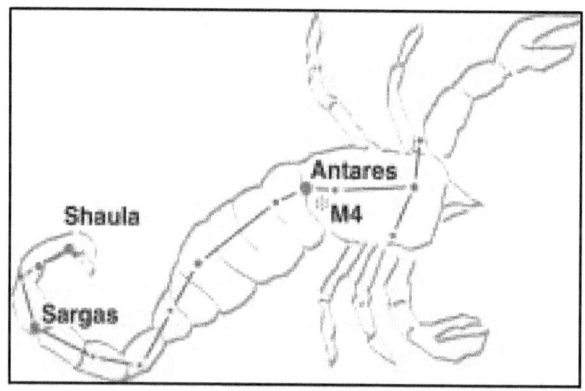

Scorpius

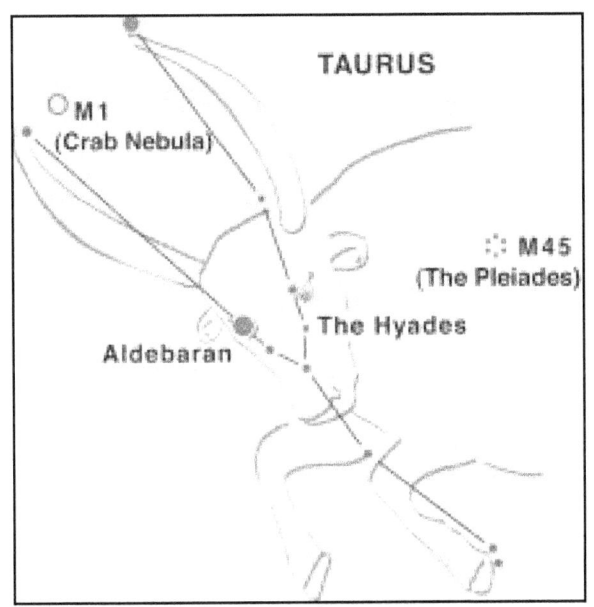

Taurus

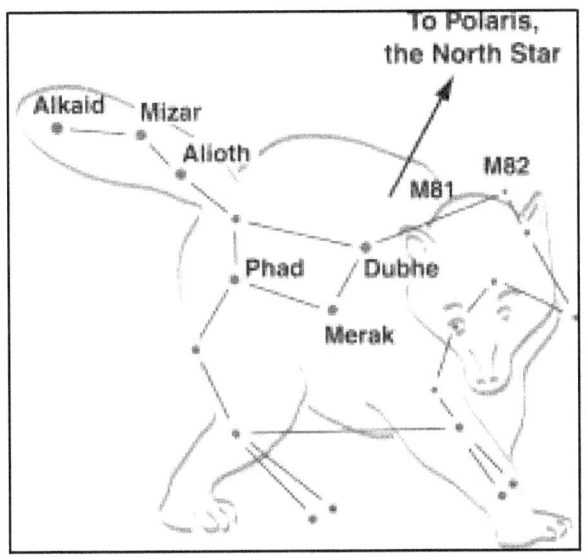

Ursa Major

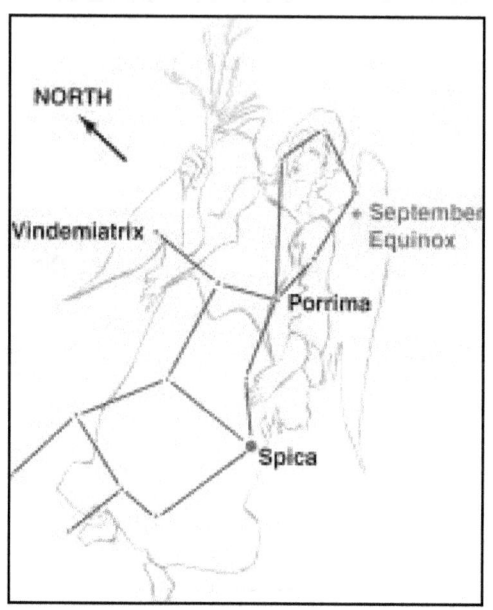

Virgo

Moon Phases

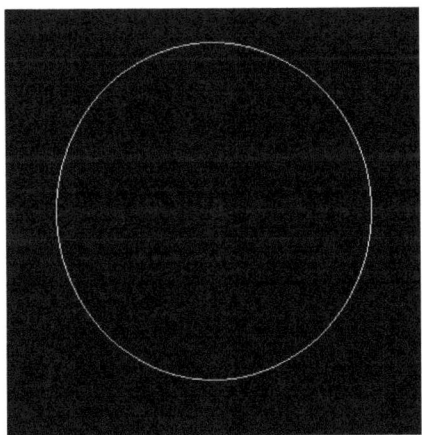

New Moon

The Moon's unilluminated side is facing the Earth. The Moon is not visible (except during a solar eclipse).

Waxing Crescent

The Moon appears to be partly but less than one-half illuminated by direct sunlight. The fraction of the Moon's disk that is illuminated is increasing.

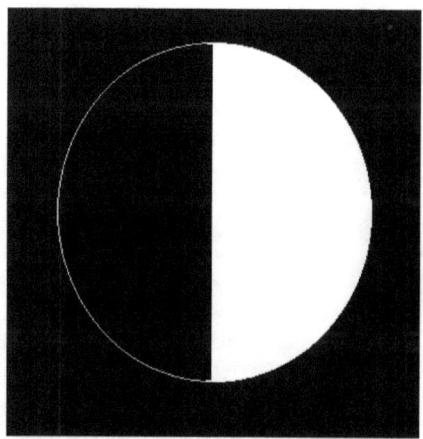

First Quarter

One-half of the Moon appears to be illuminated by direct sunlight. The fraction of the Moon's disk that is illuminated is increasing.

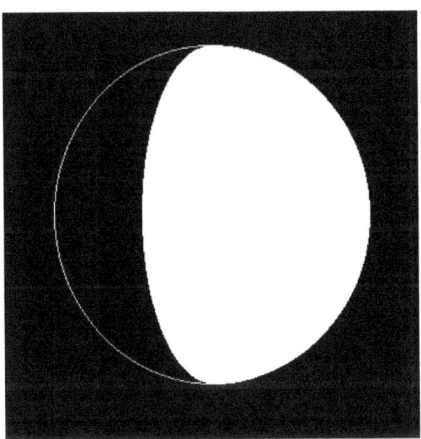

Waxing Gibbous

The Moon appears to be more than one-half but not fully illuminated by direct sunlight. The fraction of the Moon's disk that is illuminated is increasing.

Full Moon

The Moon's illuminated side is facing the Earth. The Moon appears to be completely illuminated by direct sunlight.

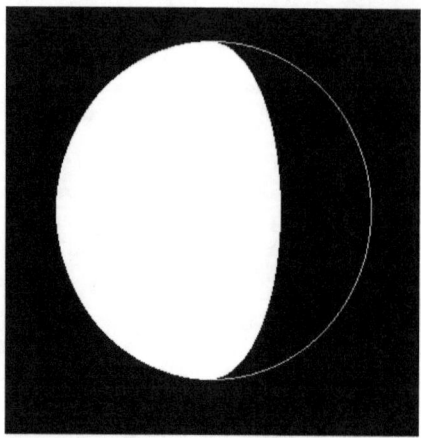

Waning Gibbous

The Moon appears to be more than one-half but not fully illuminated by direct sunlight. The fraction of the Moon's disk that is illuminated is decreasing.

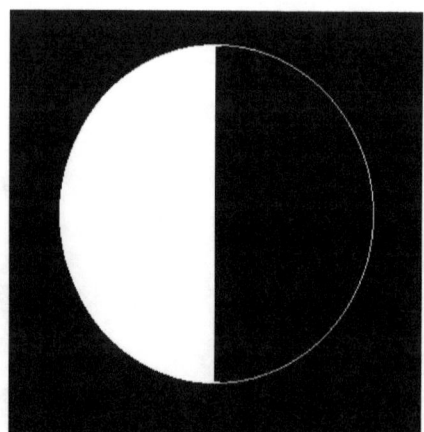

Last Quarter

One-half of the Moon appears to be illuminated by direct sunlight. The fraction of the Moon's disk that is illuminated is decreasing.

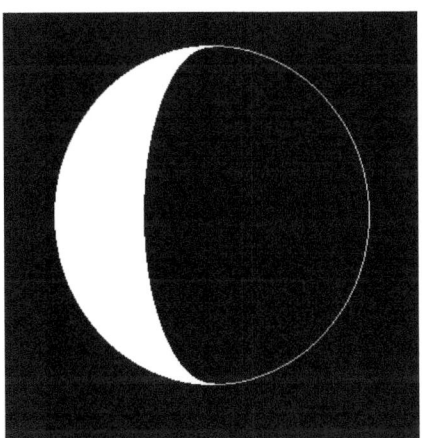

Waning Crescent

The Moon appears to be partly but less than one-half illuminated by direct sunlight. The fraction of the Moon's disk that is illuminated is decreasing.

Perspective of the Earth and the Universe

On the clearest nights, you can see over 3,000 stars. What are seen are the brightest and closest members of the Milky Way Galaxy. The remaining stars and galaxies are concealed from view by vast distances. We get a hint of our galaxy's existence when we see the misty band of light called the Milky Way on clear summer nights. The Milky Way consists of hundreds of thousands of stars too distant to register on the eye of individuals, but their combined light produces the faint misty band we see.

10,000 Mile View

Consider that we enclose the Earth, in an imaginary sphere 10,000 miles in diameter. The Earth being 8,000 miles in diameter would almost fill it. Moving outward, we expand our view of the Universe by examining what is contained in increasingly enlarged spheres, each 100 times larger in diameter than the preceding one.

1,000,000 Mile View

1,000,000 MI.

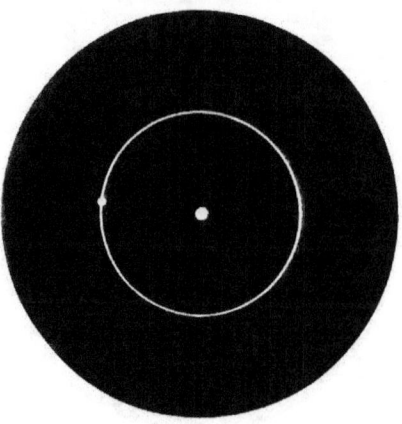

One hundred times 10,000 miles give us a sphere 1,000,000 miles in diameter, which contains the moon's orbit (one half that diameter).

100,000,000 Mile View

100,000,000 MI.

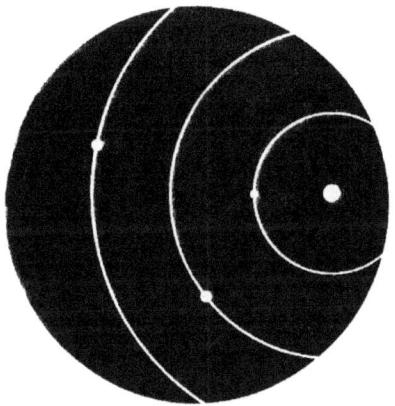

The third sphere would be 100 million miles in diameter, just a little over the Earth's distance from the Sun. The sphere includes portions of the orbits of Earth, Venus and Mercury.

10 Billion Mile View

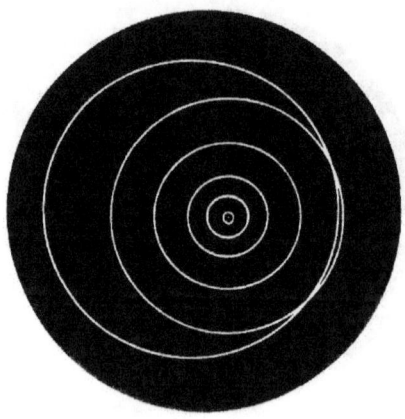

10 BILLION MI.

To include the planets in the solar system out to Pluto, we must step the progression to a sphere ten billion miles in diameter. This sphere includes all major components of the solar system.

1 Trillion Mile View

1 TRILLION MI.

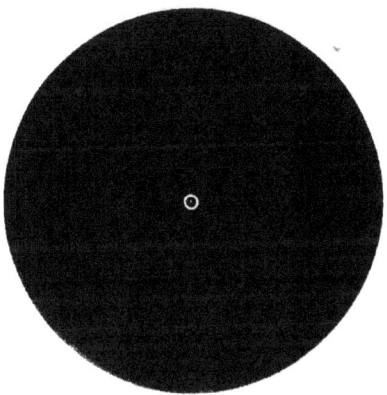

The next increase is a sphere one trillion miles in diameter. It is almost empty. Pluto's orbit shrunk to a tiny circle near the central sun, which when viewed from the edge of the sphere, would appear as simply a very bright star.

100 Trillion Mile (20 Light-Year) View

100 TRILLION MI.
OR ABOUT 20 LIGHT-YEARS

The next step is 100 trillion miles. Within this area of space are seven other stars besides the sun. Alpha Centuari is the closest, Barnard's Star is second and Sirius the brightest star in Earth's sky. At this stage, miles become an enormously awkward measuring unit. Astronomers use light- years (the distance that light travels in a year at its constant velocity of 670 million miles per hour). One light year is about 6 trillion miles. A 100 trillion diameter sphere would be about 20 light-years wide.

2000 Light-Year View

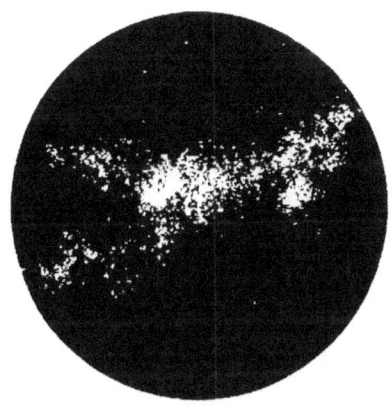

2000 LIGHT-YEARS

We make the next jump to a 2000 light-year diameter sphere. The scene becomes an ocean of stars. Stars average 6 or 7 light-years apart in the region of space around the sun.

200,000 Light-Year View

200,000 LIGHT-YEARS

The next jump to a 200,000 light-years diameter sphere shows the limits of the Milky Way Galaxy. The Milky Way Galaxy is about 100,000 light years in diameter and contains about 200 billion stars.

20 Million Light-Year View

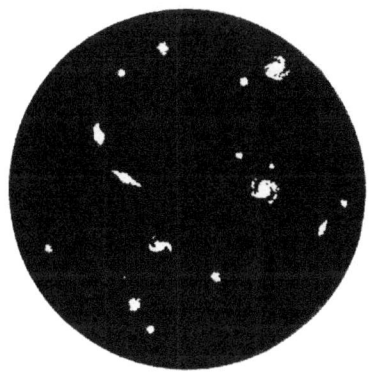

20 MILLION LIGHT-YEARS

The next step takes us to a 20 million light-years diameter sphere, where we find that our galaxy in not alone but simply one of several dozen galaxies in this volume of space. Our closest Galaxy is the Andromeda Galaxy, which is 2.4 million light years away.

2 billion Light-Year View

2 BILLION LIGHT-YEARS

The final step we will take is a 2 billion light-years diameter sphere. Within this volume, millions of galaxies exist. The Milky Way is almost lost among them.

There is little known much beyond this point. There are distant galaxies and quasars believed to be about 13 billion light-years away. Perhaps the universe is like time - that the future, folds upon itself in an elliptical form.

Part 2 - Photography Basics

Photography Basics

The Direction of Light

When it comes to the direction of light, there are 360 degrees of possibilities. When the light isn't working for you, change it by moving your position, your subject's position, or the light itself, if possible.

High Front Light (sunlight)

We are trained early on that high front light is the best type of light, and often it is.

Pros:
Most of the scene is well lit.
Bright sunny days bring out the colors of a scene.

Cons:
Sunlight may cause your subjects to squint.
Very high sunlight (seen at noon) will create deep shadows under eyes and chins, unless you use fill flash.

Front light

Front lighting illuminates the portion of the subject facing the photographer. Your camera's flash is the most common type of front lighting.

Pros:
Provides the most information to the camera by lighting the entire scene.
Easiest type of light to deal with photographically because there are fewer shadows to confuse the camera's light meter.

Cons:
Can be a bit boring—pictures lack volume and depth.
Textures and details are minimized. Scenes appear flat with few shadows.
Flash pictures may result in very bright subject areas and very dark backgrounds, if the background is beyond flash range.

Side light

Side lighting is perfect when you want to emphasize texture, dimension, shapes, or patterns. Side lighting sculpts a subject, revealing contours and textures. Use side lighting to exaggerate dimension and depth. At a 45-degree angle to the side, it's one of the most flattering types of portrait lighting.

Pros:
Can separate the subject from the background.
Conveys depth, as in a landscape at sunset.
Conveys texture, as in a weathered tree, fence, or plowed field.

Cons:
May be too severe for some subjects, creating some areas that are too bright, and some that are too dark. (See Fill flash to compensate.)

Back light

Light that comes from behind your subject is by far the trickiest to use, but the dramatic results may be worth the effort.

Pros:
Simplifies a complicated scene by emphasizing the subject, as in a silhouette.
Provides a flattering halo of light in portraits.
Adds strong shadows in landscapes.

Cons:
Lack of detail in a dark subject.
Causes lens flare resulting in low contrast and strange light spots across the picture.
Using exposure compensation to overcome backlighting results in a bright background.

Types of Light

Natural Light

The middle of a sunny day isn't the best time to take a picture.

Any kind of weather is suitable for picture-taking, and the worst weather may actually suit your subject best.

An overcast day is actually preferable for portraits - there are no harsh shadows under eyes, noses, and chins, and nobody has to squint. Flowers also photograph best on a cloudy day, especially pastel-colored flowers with soft textures.

Look at how a wet street shines and reflects headlights and traffic signals. A calm, rainy day means better reflections of the autumn leaves across the lake. In the dark fog or rain, use +0.5 or +1.0 exposure compensation.

The weather will affect the mood of your picture. Soft, foggy light will convey a very different feeling from the one the same scene conveys on a bright sunny day.

Time of Day

You may not have the luxury of waiting hours, even days, for the perfect light. Children walk away, and the tour bus has a schedule to keep. When something about a scene isn't quite right, though, consider when it would be better.

A city skyline is boring at noon. Try sunrise or sunset.

Wait until late afternoon (side lighting) to emphasize texture, like on a weathered fence.

Wait for a calm day to capture reflections on water.

Use the warm glow of late afternoon to create a romantic mood.

Other light

There are many other types of light, and most of them aren't very conducive to photography.

Fluorescent lights create a green hue. Incandescent lights, also known as tungsten (traditional light bulbs) can make a subject look yellow. And gymnasiums, stadiums, and even our streets are lit with bulbs that result in an orange or blue tint. While our eyes and brains adjust to these lights, cameras aren't that smart. To overcome these less-than-optimal situations, try one of these solutions:

Set the "white balance" feature on your digital camera to tungsten or fluorescent.

Turn on the flash (if close enough) so it becomes your main light source.

Use a high-speed film. They are generally more tolerant of mixed lighting conditions.

Use a specialty tungsten-balanced film (remember to use the whole roll).

If you can't use these techniques, take the picture anyway. If you ask, your photofinisher may be able to remove some or all of the offensive color. If you are taking digital pictures, you can easily adjust the overall color and print the pictures on your inkjet printer.

Flash, Fill Flash, and Flash Off

An automatic flash is included on just about every camera sold today. And most include a fill-flash setting for those less-than-perfect lighting situations that need a little boost. That doesn't mean the camera is fail-proof. You still need to know how and when to use these features.

General Flash Tips

Stay within flash range. Check your camera manual for the recommended range (usually 4 to 50 feet).

A higher-speed (ISO) may extend your flash by a few feet, so it does pay to use the higher-speed film, even indoors for flash pictures. With digital cameras, ISO can be changed by turning a dial.

Batteries that are approaching exhaustion will not give full flash power even if the camera is still working.

Prevent red eye by asking your subjects to look slightly away from the camera, and turn on all the room lights to shrink their pupils.

Avoid use of the "red eye reduction" flash setting as it is distracting and confusing.

Use a flash bracket to elevate the flash so that the shadow produced by the flash is lower than the subject. A bracket will also eliminate red eye. Some flash brackets provide for the use of an umbrella, which softens and distributes the lighting decreasing shadows and hot spots.

Fill Flash

Fill flash is included on most of today's cameras, and is a favorite feature. It is just enough flash to fill in areas of a picture that would otherwise be too dark.

Use fill flash for sunny day portraits to fill in those dark shadows under the eyes, nose, or under the rim of a baseball hat. It can even help in a difficult lighting situation, such as a dark complexion on a beach, or a child playing in the snow.

Fill flash is also useful for side-lit and back-lit pictures. For instance, a backlit scene may have enough bright areas in the background to provide an "average" brightness for the entire picture, but the actual subject is left in the dark. Fill flash balances the scene so that the subject is properly exposed, and the background is left alone.

Flash off

There are occasions when your camera thinks the flash is needed, but in fact it isn't. You probably have a "Flash Off" (or similar wording) setting on your camera. Here are a few examples of when to use it:

When you are too far away from your subject for the flash to be effective.

When the flash would create annoying reflections from mirrors and other shiny surfaces.

At sunset or in other low-light situations where you'd like a foreground subject to be silhouetted.

Where the quality of the existing light is beautiful, like a kitten sleeping in the sunbeam.

Where flash is not allowed (steady yourself against a wall and anchor your elbows at your side).

Quality of Light

The quality of light affects the mood of the picture.

Hard Light

Hard light, like that found on a bright, sunny day, creates very bright and very dark areas in the same scene. Another example of hard light is when the camera's flash is the only light source, resulting in bright subjects against a very dark background. Use the dark shadows as design elements or soften them with fill flash if you're within range.

Soft Light

Soft light is very camera-friendly - smooth, diffuse, even, with few shadows to confuse your camera. Cloudy days and large shaded areas offer soft light with no harsh shadows or intense bright spots.

Light even has colors. Early or late in the day, sunlight has a warm golden glow. Frigid temperatures in a snow-covered landscape can be conveyed with bluish noon-hour light.

Color Temperature

Imagine a clear incandescent lamp connected to a dimmer. As the dimmer is turned up, the voltage increases and the lamp's filament becomes warmer and warmer until it begins to glow cherry red. As the voltage continues to increase the filament gets hotter and hotter, glows more and more brightly, and is less and less red.

A blackbody is an imaginary perfect emitter and absorber of radiation. Most light sources emit light that is a mixture of light with different wavelengths. The light from a blackbody is a mixture of light with a continuous range of wavelengths. However, the light at some particular wavelength is often strongest, thus a particular color is seen. As the lamp filament gets hotter, the peak shifts toward shorter wavelengths or more towards violet.

Low color temperature implies warmer (more yellow/red) light while high color temperature implies a colder (more blue) light. Daylight has a rather low color temperature near dawn, and a higher one during the day. Therefore it can be useful to install an electrical lighting system that can supply cooler light to supplement daylight when needed, and fill in with warmer light at night. This also correlates with human feelings towards the warm colors of light coming from candles or an open fireplace at night. The standard unit for color temperature is Kelvin (k).

Color Temperature Typical Sources

1000k	Candles, Oil lamps
2000k	Very early sunrise; low effect tungsten lamps
2500k	Household light bulbs
3000k	Studio lights, photo floods
4000k	Clear flashbulbs
5000k	Typical daylight: electronic flash
5500k	The sun at noon
6000k	Bright sunshine with a clear sky
7000k	Slightly overcast sky
8000k	Hazy sky
9000k	Open shade on a clear day
10,000k	Heavily overcast sky
11,000k	Sunless blue skies
20,000+k	Open shade in mountains on a clear day

Daylight balanced film is normally, 5500k

Type A and B tungsten balanced film 3400k and 3200k

Type BCA flood lamps, 250 watts, 4,800k

Standard 100 watt incandescent bulb, 1,580 lumens, 2,850k color temp

27 watt compact fluorescent, 1,750 lumens, various color temperatures

42 watt, compact fluorescent, 2,800 lumens, various color temperatures

Basic Lighting 1

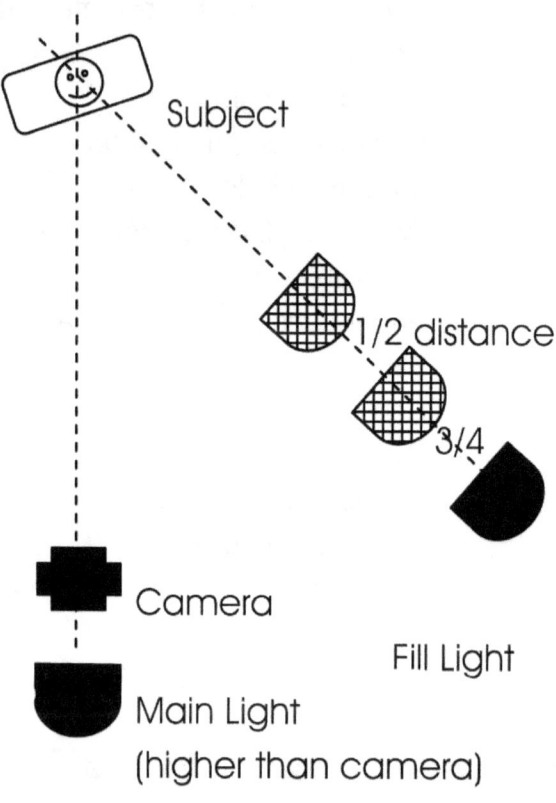

Subject

1/2 distance

3/4

Camera

Fill Light

Main Light
(higher than camera)

Basic Lighting 2

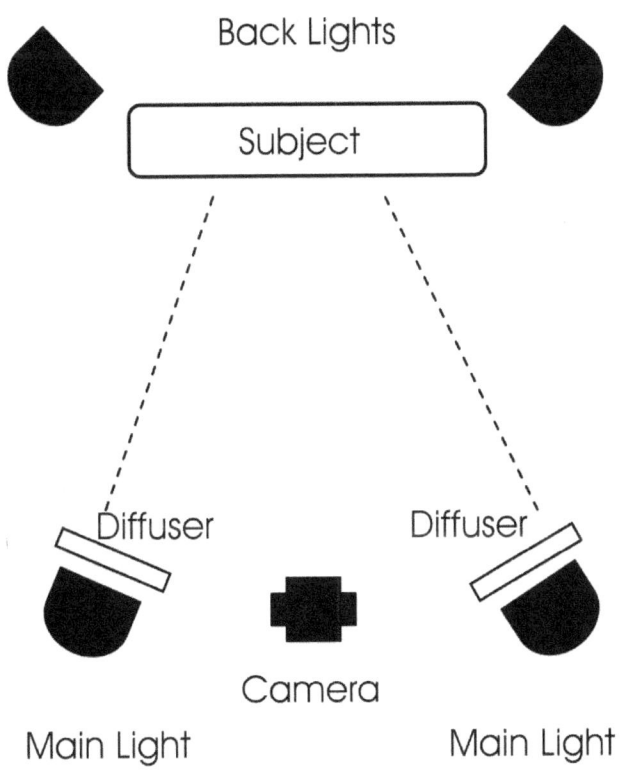

Composing Pictures

Shooting Vertical or Horizontal

You can turn your camera sideways to take a vertical picture. Hold the camera vertically to take pictures of tall buildings, waterfalls, or a person; hold the camera horizontally for groups of people, cars, and dachshunds.

Switch it Around

Try taking both horizontal and vertical pictures of the same subject to see the different effects. A subject that your might usually think of as horizontal can make a stunning vertical picture.

Although you know what your subject is, it can be hard for a viewer to determine your intent if too many elements in your picture make it confusing. Eliminate all unimportant elements by moving closer, zooming in, or choosing a different shooting angle.

For the most complimentary portrait, shoot at your subject's eye level. However, if you want to have some creative fun, change your angle of view.

Alter Your Position

Change your position to emphasize or exaggerate how big or small your subject is. Crouch down and shoot up at someone and that person towers over you. Shoot down on your pet and it seems so comically small. You can also move your camera right or left only a few feet to change the composition dramatically.

Placing the Subject off Center

Putting the subject off-center often makes the composition more dynamic and interesting. Even if your subject fills the frame, the most important part of the subject (for example, the eyes in a portrait) should not be dead center

Follow the rule of thirds. An easy way to compose off-center pictures is to imagine a tick-tack-toe board over your viewfinder. Avoid placing your subject in that center square, and you have followed the rule of thirds. Try to place your subject along one of the imaginary lines that divides your frame.

Watch the horizon. Just as an off-center subject is usually best, so is an off-center – straight-horizon line. Avoid cutting your picture in half by placing the horizon in the middle of the picture. To accent spaciousness, keep the horizon low in the picture. To suggest closeness, position the horizon high in your picture.

Using Leading Lines

Select a camera angle where the natural lines of the scene lead the viewers' eyes into the picture and toward your main center of interest. You can find such a line in a road, a fence, even a shadow. Diagonal lines are dynamic; curved lines are flowing and graceful. You can often find the right line by moving around and choosing an appropriate angle.

Avoiding Distracting Backgrounds

Select an uncomplicated background that does not compete with your subject. Bright colors and text (for example, store signs) create the biggest problems. Be especially aware of what is behind your subject in a portrait so that branches don't accidentally become antlers.

Move your subject or change your camera angle to find a simple, uncluttered background. Taking this extra step before you press the shutter button makes a big difference in the end result.

Including Objects in the Foreground

When taking pictures of landscapes, include an object, such as a tree or boulder, in the foreground. Elements in the foreground add a sense of depth to the picture. A person in the foreground helps establish a sense of scale.

Sometimes you can use the foreground elements to "frame" your subject. Overhanging tree branches, a doorway, or an arch can give a picture the depth it needs to make it more than just another snapshot.

Photographing Waterfalls

Water is life. The existence of nearly everything living is critically linked to its availability. In and near the streams, rivers and oceans of our world, Nature thrives. It is no surprise that such a powerful force in Nature is so often depicted by the outdoor photographer.

Use slow shutter speeds to create a soft, artistic portrait of water. Generally, shutter speeds that are 1/6 of a sec and slower will yield the best results. A majority of waterfall photographs fall between 1/4 sec and 3 seconds of exposure. The key here is to dare to experiment and not be afraid to shoot a lot of film. You never know what shutter speed is going to render the waterfall the way you see it in your mind's eye. Needless to say, you'll need a tripod when making these long exposures.

When shooting waterfalls because they're usually reflecting a lot of light that is going to fool your camera's built-in exposure meter unless you compensate. Spot metering off of something neutral in the same area (and thus in the same light) works best; rocks, tree trunks and grass are usually good candidates.

Use a circular polarizer and/or neutral density filter. The polarizer can increase the overall color saturation in the scene as well as decrease your shutter speeds by 1.5 to 2 stops (a good thing if you find yourself battling a rising sun with an overall increase in lighting). Slowly rotate the polarizer to witness the effect it is having on your scene. Depending on the lighting conditions and your position relative to the sun, you may or may not decide to use it. A neutral density filter can also be used to reduce light input by up to 2 stops in most cases (depending on the strength of the filter).

Using a slow speed film will enable you to photograph waterfalls with a variety of long exposures. Slow speed films will also reward you with incredibly fine grain and outstanding color saturation.

Camera Basics

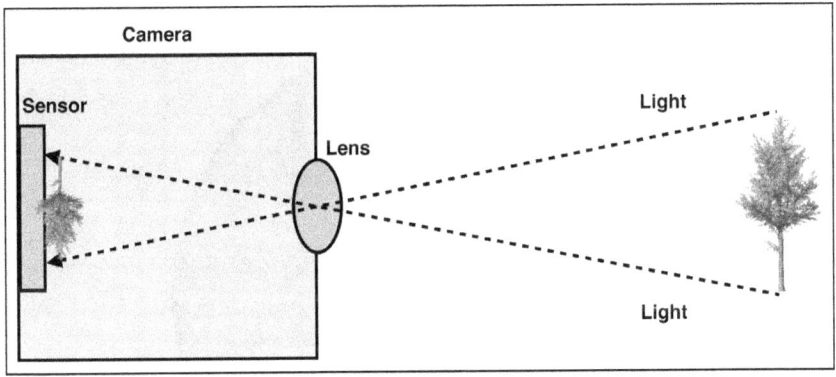

Light enters a camera through a lens. The lens is moved forward and backward to focus the light onto the sensor or film. A pinhole can replace a lens. It is that simple.

Well, it really is not that simple. There are stumbling blocks like lens distortions. Spherical aberrations can cause the image to blur at the edges. Chromatic aberrations can cause the colors to be distorted. Consequently, lenses are made of many different elements and coatings to solve these problems. A good quality lens may have more than a dozen elements.

Aperture

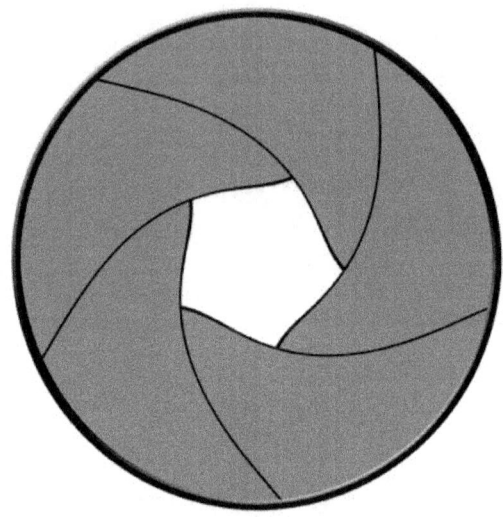

The camera's aperture controls the amount of light, which reaches the film or sensor. This function plays a vital role in one of the basic aspects of photography: depth-of-field.

Large apertures (f/2.8-f/5.6): produce shallow depth of field. This means the area of sharp focus in the picture will be small. This can be useful when you want to isolate the subject of your picture while throwing the background and other distracting elements out of focus. Some useful applications of wide apertures include portraits and wildlife close-ups.

Small apertures (f/16-f/32): increase depth of field, which means more elements of a picture, from foreground to background become sharply focused. This can create a distinct sense of depth to a photograph, drawing the viewer into the picture. Small apertures are essential for most landscape photographs.

Lens "sweet spots" (f/8 & f/11): mid-range apertures of f/8 & f/11 due to technical aspects of the lens optics, often yield the sharpest images. When neither a large or very small aperture is needed, these are good apertures to use to maximize the sharpness the lens can deliver.

Telephoto

High shutter speeds are necessary to get sharp pictures with a telephoto. When an image is magnified, you also magnify the effect of camera movement. You can counteract this by using a fast shutter speed. This gives any slight movement little chance to blur your image. A good rule of thumb is to shoot at a shutter speed of 1/focal length or faster. For example, if your zoom lens is set at 200 mm, shoot at a shutter speed of 1/250 or greater.

High speed film or sensor (ISO 400+) is more sensitive to light. That means it needs less light to get the same exposure (compared to say a 100 or 200 speed). Less light amounts to higher shutter speeds that freeze both camera AND subject movement.

Use a wide aperture when you want to isolate your subject and throw a distracting background out of focus. Apertures like f/5.6 or wider will usually do the job, especially if your subject is located a good distance away from the actual background. If your subject is a person or animal, focus on the eyes. It is often better to frame the subject against a dark colored background in order to capture the greatest degree of detail in your subject.

Use a tripod whenever you can. Nothing beats a strong, heavy tripod in delivering sharp telephoto shots. Invest in a high quality tripod and your pictures will show it!

Maximizing Sharpness

Use a slow ISO (25-100 speed) whenever possible. The lower the ISO number, the sharper the image. Keep your subject in mind, however, when making this decision. You won't benefit from the films sharpness if your shutter speeds are too slow to stop subject movement.

Use a medium or high-speed ISO when using slow (f/4.5-5.6) telephoto lenses, especially if you'll be handholding the camera.

Manual focus becomes critical when your zoom lens is at a maximum setting at a wide open aperture (ie: f/2.8-5.6) - the depth of field will be so shallow that its critical to do the fine tuning yourself to insure that the important elements in your photograph are crisp such as an animal's eyes.

Camera support is synonymous to sharp, highly detailed photographs. A firm, heavy tripod is best but may not always be practical. ANY form of camera support is better than nothing - whether you're using a picnic table to steady your arms, a clamp, monopod, or mini-tripod.

Use the "sweet spots" whenever possible - midrange apertures (f/8-f/11) and zoom settings (150 mm on a 80-300 mm lens) make the most of your lens optics, resulting in sharper photographs.

Exposure

Most cameras in use today have a built-in light meter that examines the light reflected by the subject you're framing in the viewfinder. Understanding how this light meter reads the scene is crucial to giving you the knowledge you need to make successful, well-exposed photographs the majority of the time you shoot. Or, you can simply set the camera on automatic exposure.

Use exposure compensation. Most cameras have this feature, which allows you to adjust the camera's built-in light meter reading. For subjects lighter than average "add" exposure by setting the dial to plus 1/2 or plus 1 stop (or more depending on the subject). Darker than average subjects require a minus setting (some cameras have a chart that goes from -2 stops to +2 stops with intervals of a 1/2 or 1/3 stop).

Bracket. Extra insurance for tricky lighting situations and subjects. Bracketing your exposure means shooting over and under the exposure setting recommended by the camera (or the one you've set manually). This can be crucial when shooting slide film where even minor exposure errors can result in an unusable image.

Depth of Field

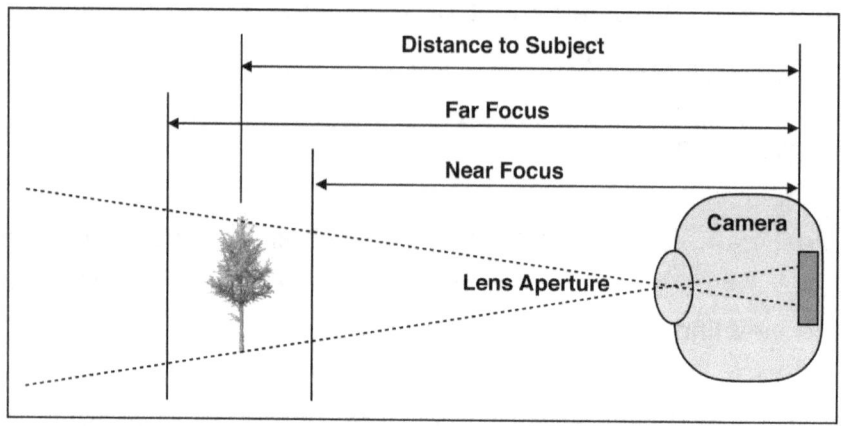

Lens (mm)	f number	Distance (feet)	Near Focus (feet)	Far Focus (feet)	Difference (feet)
50	2	10	9.54	10.50	0.96
50	5.6	10	8.81	11.60	2.76
50	2	20	18.20	22.10	3.91
50	5.6	20	15.70	27.50	11.8

Nikon D7000

Depth of field (DOF) is the sharpness of objects(s) at a specific range of distances. Sometimes, such as when shooting scenery, everything should be in perfect sharpness. However, sometimes, such as when shooting a flower, the flower should be sharp and the background un-sharp. The smaller the aperture, the greater the depth of field. The greater the distance, the greater the depth of field. The greater the depth of field, the greater the sharpness range. Points located before or after the focus plane will not show as points, but as circles, called circles of confusion.

The hyperfocal distance is the near distance of the object(s) that are in focus when the lens is focused to infinity. When the focus is set to the hyperfocal, all the planes from half the hyperfocal distance to the infinity will be in focus.

(Hyperfocal distance) $H = (F \times F) / (A \times C)$

Where:
C: diameter of the circle of confusion (CoC)
H: Hyperfocal distance
F: lens Focal length
A: lens Aperture (f stop)

Circle of confusion Table

Film format	**Frame size**	**CoC**
Small Format		
Four Thirds System	18 mm × 13.5 mm	0.015 mm
APS-C	22.5 mm × 15.0 mm	0.018 mm
35 mm	36 mm × 24 mm	0.029 mm
Medium Format		
645 (6×4.5)	56 mm × 42 mm	0.047 mm
6×6	56 mm × 56 mm	0.053 mm
6×7	56 mm × 69 mm	0.059 mm
6×9	56 mm × 84 mm	0.067 mm
6×12	56 mm × 112 mm	0.083 mm
6×17	56 mm × 168 mm	0.12 mm
Large Format		
4×5	102 mm × 127 mm	0.11 mm
5×7	127 mm × 178 mm	0.15 mm
8×10	203 mm × 254 mm	0.22 mm

Depth of Field is calculated as follows:

$$\text{(Near focus plane) } NF = (H \times D) / (H + (D-F))$$

$$\text{(Far focus plane) } FF = (H \times D) / (H - (D-F))$$

Where:

NF: Near Focus Limit (mm)

H: Hyperfocal Distance (mm)

D: Subject distance (mm)

F: lens Focal length

Multiply Inches by 25.4 to convert to Millimeters. Divide Millimeters by 25.4 to convert to Inches.

As the aperture increases, the DOF decreases.

As the focal length increases, the DOF decreases.

As the distance to subject increase, the DOF increases.

Shutter

Shutter speeds usually vary from 1/8000 second to 30 seconds.

Shutter speed steps are typically 1/250, 1/125, 1/60, 1/30, 1/15, and so on.

1/60 second is usually the limit of hand holding a camera.

Digital Photography

Cameras

Digital cameras are similar to film cameras in the way photographs are taken. In digital cameras, the film is replaced by a CCD or CMOS computer chip that captures the light that is reflected from a scene. The light is stored as red, green, and blue pixels. The files are stored on memory storage cards.

Most digital cameras have an LCD screen for viewing photographs. Viewing photographs instantaneously, allows the photographer to see the final result. If not satisfied with the picture, the photographer can shoot again.

The main advantages of digital photography are the ability to see the results instantly and the ability to edit the photographs using a computer. It's like having a darkroom in the computer or camera. The images can be imported into the computer either directly from the camera or from the storage card or seen directly on the camera view screen.

Camera resolution is measured in millions of pixels. This is the primary factor in determining image sharpness and presentation size. Larger prints can be produced using software like this that interpolates between pixels.

There is no standardization for the physical size of pixels. Consider the size of the sensor array and the number of pixels in that array. Digital Single lens reflex (DSLR) cameras have much larger sensors than the point and shoot camera.

For example, the Canon G6 point and shot camera has a 7.18 x 5.32 mm sensor. The Canon Rebel EOS 400 and Nikon D60 DSLR cameras have 22 x 14.8 mm sensors. The Canon 5D and Nikon D700 DSLR cameras have 23.9 x 36 mm (full frame) sensors. Physically larger pixels produce less noise than smaller pixels as seen in the small sensors of a point and shoot cameras. Lower noise produces cleaner images. It is actually easier to get a higher ISO from the bigger pixels because more light falls on it (bigger area) causing more photo-electrons to be produced. There comes a point where too many pixels are squeezed into a sensor array resulting in too much noise.

Basically, point and shoot cameras require more light than DSLR cameras. They have the advantage of being small and light. Most point and shoot cameras have a built in flash with a guide number of only 13 feet. If a point and shoot camera was chosen, it should be one that has a hot shoe for an external flash. Canon and Nikon make flashes that can fit in a pocket with a guide number of 75+ feet.

White Balance and Color Temperature - A primary objective in photography is to accurately replicate colors of the subject. "White balance" is a digital camera function to correct for color shifting resulting from different light sources. All light sources have a specific color temperature measured in degrees Kelvin. For example incandescent light bulbs have a temperature of about 3700K and daylight can vary depending on location and time of day from 3000K - 8000K with mid day temperature of about 5500K. One of the advantages of digital cameras over film is the ability of the camera to compensate for different color temperatures thereby allowing for more accurate color replication. Digital cameras vary widely in the methods and features to control color temperature, with more expensive cameras having superior controls. Almost all digital cameras have an "auto white balance" function, which is suitable for most situations.

Another major advantage of digital cameras is that the ISO can be adjusted. Some of the newer cameras have ISO's to 6,400 without excessive noise.

Sensor Sizes

A full frame camera projects an image onto a sensor or film that is 36 mm x 24 mm in size. An APS-C camera projects an image that is 15 mm x 22.5 mm in size. An APS-C sensor costs less than a full frame sensor.

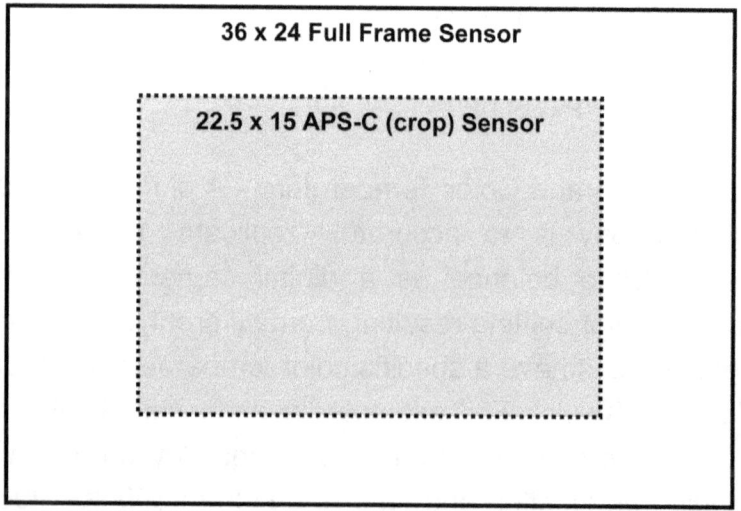

36 x 24 Full Frame Sensor

22.5 x 15 APS-C (crop) Sensor

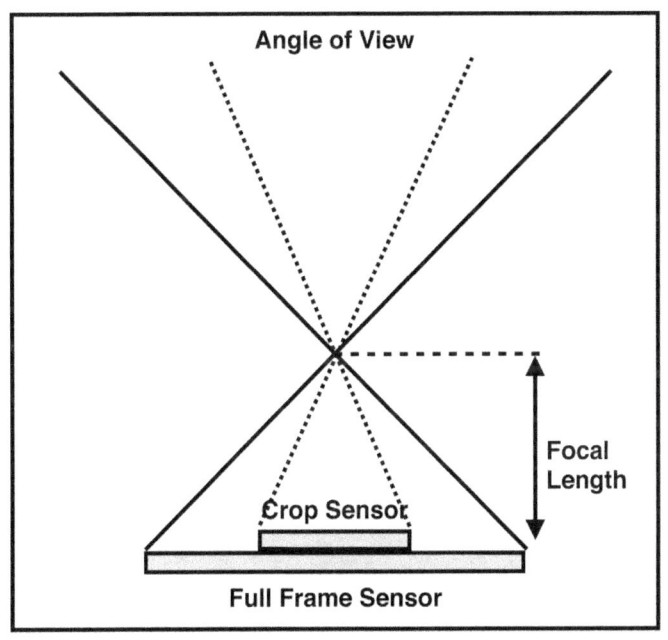

Angle of View

Focal Length

Crop Sensor

Full Frame Sensor

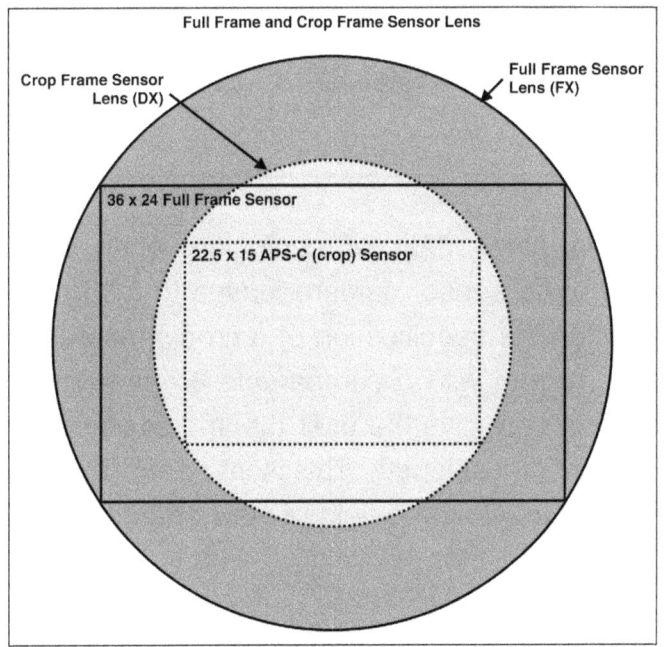

Full Frame and Crop Frame Sensor Lens

Crop Frame Sensor Lens (DX)

Full Frame Sensor Lens (FX)

36 x 24 Full Frame Sensor

22.5 x 15 APS-C (crop) Sensor

APS-C (Crop Sensor) Camera with 70 mm FX Lens (actual size)

Full Frame Sensor Camera with 70 mm FX Lens (actual size)

There is a great deal of confusion when it comes to comparing crop sensor camera lenses to full frame camera lenses. Often the specification of a crop sensor lens may be stated as 10 mm with a comparable 35 mm focal length of 15 mm. This may give the user the impression that they will have a 15 mm focal length. This is incorrect. The focal length doesn't change. Only the field of view changes. Therefore, it should be stated that the field of view is the same as a 15 mm focal length lens.

A full frame camera projects an image onto a sensor or film that is 36 mm x 24 mm in size. An APS-C camera projects an image that is 23.5 x 15.6 mm in size. An APS-C sensor costs less than a full frame sensor. APS-C lenses are smaller and lighter in weight. Larger sensors are often used for larger size prints.

What is of primary concern is the field of view that is produced by these two sensors. As seen in the diagram below, the field of view that is seen by the full frame sensor is wider than the field of view seen by the APS-C or "crop" sensor. The focal length remains the same, which is often confused. The only aspect that changes is the field of view, not the focal length.

Lens and the images projected onto a sensor are round in shape, whereas the sensors are rectangular in shape. The diameter of the circle needs to be larger than the diagonal of the rectangular sensor. A full frame 35 mm lens must have an image circle larger than 43.27 mm. An APS-C camera lens needs to have an image circle larger than 27.04 mm. If an APS-C lens is used on a full frame camera, the image circle would not be large enough to cover the corners of the sensor. If a full frame camera lens is used on an APS-C camera, it will cover the corners of the APS-C sensor.

What makes this concept confusing is that when photos are added to a photo software program, they are often enlarged by the software to a specific size. Most people don't pay attention to the enlargement size in the software. Photos are then presented online using the same frame size, which gives the appearance that they are enlarged. But, the photos are not enlarged from the camera, only from the photo software. In this article, the actual photos are presented in actual size format.

When a DX lens specification says, for example, a 100 mm DX lens is equivalent to a 150 mm FX lens, it is referring to the field of view, not the magnification. It should be stated as the field of view of a 100 mm DX lens is equivalent to the field of view of a 150 mm FX lens.

To obtain the field of view obtained with the FX camera using a DX camera, change the lens (or zoom out) from 70 mm to 46.6 mm. In order words, a photo taken with a DX camera and 46.6 mm lens would look exactly the same size as the photo taken with an 70 mm FX camera lens. The resolution of an FX camera will be greater than that of a DX camera because the full frame sensor is larger and contains more pixels per size. This is important when printing larger prints. It probably won't make much difference when displaying photos on the web or printing smaller sizes such as 8" x 10".

Lenses

The longer the focal length of the lens, the narrower is the field of view. For portraits, a distance of about 6 feet is used with a lens that has a focal length of about 100 mm. A zoom lens will allow for variations in the field of view. Using lenses of focal length greater than 100 mm will require the use of a tripod to eliminate blurring of the image from camera shake.

Image Size (mm) = Focal Length (of the lens in mm)

109

Focal Length (mm)	Field of View (degrees)
17	104
28	75
50	47
85	28
135	18
300	8
500	5
1000	2.5

File Compression Ratios

Raw vs jpeg is a highly controversial topic. However there are different formats of jpeg with different compression ratios. Compression algorithms don't simply reduce file sizes. They are complicated algorithms that take out repeated data or what seems to be repeated data in a music, video, or photographic file.

One test I performed with a Sony A6300 camera resulted in:

Standard jpeg 3.5 mb
Fine jpeg 6.6 mb
Extra fine jpeg 13.7 mb
Raw 25 mb

The above file sizes will vary depending upon the data contained within a photograph. One could roughly conclude based on file size that the extra fine jpeg is compressed approximately 2:1. The fine jpeg is compressed approximately 4:1. The standard jpeg is compressed approximately 7:1.

Backyard Bird Photography

Introduction

Bird photography is very difficult because birds move around quickly and are easily startled. Originally, the bird feeder mount was placed at a distance of 20 feet from a residence. When anyone came near the window or back door of the residence, the birds flew away. Consequently, the feeder had to be moved to a distance of 40 feet.

APS-C (crop sensor) cameras are lightweight. However, significant cropping and enlarging will significantly degrade the image. APS-C camera will require longer focal length lenses to avoid cropping and enlarging. Sample photographs are shown at various focal lengths for APS-C and Full Frame cameras. The longer the focal length, the larger the image. However, usually the longer the focal length, the heavier the lens, and the more expensive the lens is. Also, the longer the focal length, the more susceptible will be the photograph to camera shake and will usually require a monopod or tripod. An image stabilized lens will help to compensate for handheld camera shake. A moderately priced image stabilized lens that is often used for birds and wildlife is the Sigma or Tamron 150-600 mm f5-6.3 lens. It costs about $1,000 and weighs 4.3 pounds.

Teleconverters (1.4x and 2x) are sometimes used to increase the focal length of a lens. However, they increase the focal length while sacrificing aperture and image quality.

A shutter speed as fast as possible will be required to freeze the action. A minimum shutter speed of 1/500 second is recommended. A faster shutter speed is always better. Slower shutter speeds will require the use of a monopod or tripod. However, bird movement may blur the photograph at slower shutter speeds. A telephoto lens with a focal length greater than 300 mm will require a faster shutter speed. A shutter speed of 1/800 second should be adequate for most circumstances. A higher ISO will be required to obtain a faster shutter speed. Modern cameras can use high ISOs without any noticeable image degradation.

The larger the aperture, the better the bokeh. Bokeh is a term used when the image is sharp and the background is blurred. Bokeh is produced by the lens itself and the shorter depth of field. Photographing an image that is sharp with a blurred background gives the impression of depth to the photograph. An aperture of f/5.6 has a larger opening than an aperture of f/8. At 40 feet, a 600 mm lens at f/8 will provide a depth of field of -3.68" and + 3.74" on a FF camera. Apps for smartphones can be used to calculate depth of field.

As can be seen, the 300 mm lens is not quite adequate to photograph sparrows at a distance of 40 feet. A longer focal length lens of 600 mm had to be used. Getting closer to the subject would only cause them to fly away. Then, there is the position of the Sun...

For more technical information about cameras and photography read the book "Introduction to Astronomy and Photography" by Dr. John A. Allocca.

Bird Feeder Mount

The bird feeder mount shown below was constructed using a heavy duty umbrella base and a 1-1/4 inch diameter powder coated steel closet rod, which was cut down to 80 inches for this project. Four closet rod brackets were mounted using 10-32 / 2 inch stainless steel bolts and nuts though a 1/4 inch hole. A 1-1/4 inch plastic cap was inserted on the top of the pole. Mount either 2 or 4 bird feeders to maintain balance. A squirrel cone is mounted on the pole.

Ace Hardware:

Mounting Base for 1-1/4 umbrella pole: many available

John Sterling 96 inch Length x 1-1/4 inch Diameter Powder Coated Steel Closet Rod, Item no. 5209689, $14.99

Knape & Vogt John Sterling White White Steel 12 Ga. Shelf/ Rod 1 in. H x 12.4 in. W x 10 in. L, Item no. 54724, $5.99

John Sterling Pro 1-1/4 in. L x 1-1/4 in. Dia. White Plastic Closet Rod Support
Item no. 5209697, $2.99

10-32 / 2 inch stainless steel bolts and nuts

Sample Photographs

Sony A6300 APS-C, 55-210 mm at 55 mm, f/8, 1/640 sec, ISO 3200, 40 feet, Handheld

Sony A6300 APS-C, 55-210 mm at 210 mm, f/8, 1/800 sec, ISO 3200, 40 feet, Handheld

Sony A7RIII FF, 70-300 mm at 70 mm, f/8, 1/500 sec, ISO 3200, 40 feet, Handheld

Sony A7RIII FF, 70-300 mm at 100 mm, f/8, 1/500 sec, ISO 3200, 40 feet, Handheld

Sony A7RIII FF, 70-300 mm at 200 mm, f/8, 1/800 sec, ISO 3200, 40 feet, Handheld

Sony A7RIII FF, 70-300 mm at 300 mm, f/8, 1/1000 sec, ISO 3200, 40 feet, Handheld

Sony A7RIII FF, 150-600 mm at 400 mm, f/8, 1/640 sec, ISO 3200, 40 feet, Monopod

Sony A7RIII FF, 150-600 mm at 500 mm, f/8, 1/640 sec, ISO 3200, 40 feet, Monopod

Sony A7RIII FF, 150-600 mm at 600 mm, f/8, 1/500 sec, ISO 3200, 40 feet, Monopod

The image above was cropped 1/7th from the 210 mm photo taken with the A6300

The image above was cropped 1/6th from the 300 mm photo taken with the A7RIII

The image above was cropped 1/4th from the 600 mm photo taken with the A7RIII

Notice the higher resolution in the photograph above because the crop was not as significant (1/4th) using the 600 mm lens than the shorter 210 mm (1/6th) and 300 mm (1/7th) focal length lenses.

Monopod

The following monopod was used for this project:

Manfrotto Xpro Monopod Plus Carbon Fiber 4 Section
Min Height 20.47 in
Maximum Height 64.76 in
Legs Tube Diameter 1.15- 0.98- 0.8- 0.63 in
Safety Payload Weight 15.43 lbs.
Leg Sections 4
Top Attachment 1/4" screw, 3/8" screw"
Weight 1.3 pounds, $220

XPRO Ball Head in magnesium with 200PL plate
MHXPRO-BHQ2
Load capacity 22 lbs.
200PL-14
Weight 1.1 pounds, $130

Bird Feeders and Supplies

Wagner's 62059 Greatest Variety Blend, 16-Pound Bag
A Blend that attracts the greatest variety of Songbirds.
Loaded with Sunflower seeds, Songbirds' favorite
11 ingredients - A superb blend
For tube, hopper, or platform feeders
Resealable Velcro® closure
Attracts perching and ground Songbirds
Satisfaction guaranteed
Available sizes: 6 & 16 lb. bags
The blend is packed with Black Oil Sunflowers, Striped Sunflowers and Sunflower Chips which are the top food choice for the greatest number of wild bird species.
By offering Sunflower seeds in different sizes, both hulled and unhulled, birds of various sizes can find their favorite food.
White Millet, Red Millet, Cracked Corn, and Red Milo will attract a variety of ground feeders such as Tree Sparrows, Song Sparrows, House Sparrows, Dark-eyed Juncos, and Mourning Doves to your backyard.
Nyjer, Peanut Kernels, Canary seed and Safflower will attract specific species of preferred birds ranging from Cardinals, Chickadees and Titmice to Jays, Finches and Woodpeckers.
$19.99

Wide Deluxe Easy Clean Tube Feeder (Model# WMRS-18)
Natures Way Birds
Easy Clean™ technology - simply lift and slide ports and base to remove and clean
Even Feed™ baffles allow for constant seed and birds at all levels

Unique hanger design connects to outside of tube, removing any obstacles for the easiest filling and less seed spilling
4" wide tube enables easier filling with less seed spilling
Durable powder-coated metal roof and ports
2-in-1 seed capability; Thistle inserts included!
Stay-clear, break resistant tube
4 feeding ports
Limited Lifetime Warranty
Dimensions: 17"H x 7"W x 7"D   Weight: 1.5 lb Capacity: 3qts

Seed to use: Black oil sunflower, striped sunflower, hulled sunflower, safflower, cracked corn, nuts, mixed; thistle inserts for feeding nyjer
Attracts these birds: Cardinals, Grosbeaks, Titmice, Nuthatches, Chickadees, Juncos, Finches
$19.95

North States Village Collection Around Town-Birdfeeder-School House
Charming school house-themed bird feeder is made of plastic for easy cleaning
Cord attached for easy hanging
Feeder top lifts to fill
Holds up to 5-Pound of seeds
13-1/2-Inch by 9-Inch
$17.95

North States Village Collection Around Town-Birdfeeder-Log Cabin

Complete with windows and doors, the solid resin construction of this log cabin feeder will keep your feathered neighbors happy

Lift removable chimney to fill with up to 8-Pound of seed
Can be used as a hanging feeder or pole mounted
3 year guarantee
14-Inch by 10-Inch by 11-Inch
Seed Capacity: 5 lbs.
$19.95

Woodlink NASCOOP Audubon Seed Scoop
Audubon seed scoop
Made of plastic
Narrow handle at the base for a comfortable grip
A wide mouth with a spout for easy pouring
Last for several seasons
$10.80

Woodlink NABAF18 Audubon Wrap Around Squirrel Baffle, 18-Inch
Made of textured powder coated steel
Defeats squirrels every time
No need to remove bird feeder when mounting
Secure black coupler (included) around any 1/2-inch to 1-3/8-inch pole or shepherd hook
Snaps apart with inward pressure on each side of the seam. Additional pieces that will hold the cover are tapped to the bottom of unit.
$18.95

Part 3 - Astrophotography and Observing

Astrophotography Basics

Using live view and digital astrophotography is helpful for people with poor night vision and fun for those with normal night vision. These cameras can often capture more detail than the human eye can. The celestial view can be observed on a monitor or computer by a group of people simultaneously. One can also connect the camera to a wifi enabled telescope and run the entire observation session from indoors after setting up the telescope.

There are many different methods for live view and digital astrophotography. One method is to use a digital or film single lens reflex camera (DSLR for SLR) with an adapter. DSLR and SLR astrophotography are advanced methods, which will only be briefly discussed here. Next, there are a number of analog CCD and digital cameras on the market specifically designed for telescopes and microscopes. The CCD cameras adapted for telescopes usually come with a 1.25" eyepiece and focal adapter. One simply replaces the eyepiece from the telescope with the camera. That was easy. There are two basic methods of connecting the camera. One is to connect a digital camera with a USB connection to a computer and the other is to connect an analog CCD camera to an analog video monitor. Both methods can be used to view images in real time. Currently, analog CCD cameras, similar to those used for security cameras, have a much higher resolution than the digital cameras. Digital cameras with 5 megapixel resolution is only fair in resolution. As of this writing, Celestron is coming out with a 10 megapixel (mp) digital camera for $250. Research grade microscopes use either an analog CCD camera or a 20 mp digital camera for $5,000. Computers and software usually have more advanced features, and are often more complicated to use. The analog CCD camera to analog monitor method is the simplest and easiest to use.

The light from the moon is bright. But, from the stars, the light is dim. Too dim for normal photography. Dim light from the stars requires long time exposures, often several hours. With digital cameras and software that are designed for telescopes the cameras and software take continuous exposures so that they can be stacked one on top of another. This is similar to long time exposures with a DSLR or SLR gathering light over time. The result is a high resolution photograph. If a long time exposure is used, a tracking tripod or GOTO telescope will be required to track the path of the stars during recording. Video can also be recorded with these cameras.

Celestron, Meade, and other manufacturers offer cameras ($100-$300) that plug into a computer via a USB cable. The open source oacapture software available from http:// www.openastroproject.org will work with a Mac. As of the date of this printing, the Troupe software does NOT work with a Mac and USB camera. The Revolution 2 system ($300), available from http://www.revolutionimager.com, is an easy to use complete analog setup designed to be used with a portable monitor and rechargeable battery.

Most software that is available to be used with USB digital cameras are primarily designed for the PC. The following software was tested on a Mac on 8/29/16.

Computer: iMac (21.5-inch, Late 2013)
Operating System: OS X 10.11.6 El Capitan

Camera used to test software: Neewer USB 640*480 Pixel 5.0 megapixel Webcam Camera

FaceTime: Worked fine.

Astroimager, Version 2.3: Did not recognize webcam camera.

Lynkeos, Version 2.10: Did not recognize webcam camera.

Meade-Sky-Capture, Version: MacOS, 3.7.1: Did not recognize webcam camera. This software works with Meade cameras only. It was tested on the Meade LPI-G color camera.

oaCapture, Version 1.0.0: Recognized webcam camera. Appears to work fine.

Starlight Live, Version 3.1: Did not recognize webcam camera.

DSLR Astrophotography

Introduction

Astrophotography can be rewarding and yet frustrating at times. Purchasing the right equipment for the desired effect may be challenging. Locating clear dark skies can also be a challenge. Equipment may need to be portable enough to be transported to a clear dark sky location. Computer simulations of equipment is a great way to try equipment out before making a purchase. Asking for advice is always a good idea. Canon cameras are generally more friendly towards astrophotography because they began the art. This author used Nikon because that was purchased previously for other uses.

Earth's Rotation

As the earth rotates the stars appear to rotate through the sky above. By aligning the telescope to a fixed point in the sky which isn't moving allows one to track objects using only the Right Ascension (RA) control. The Right Ascension movement compensates for the earths movement and allows the telescope to track an object. The part of the sky that doesn't move is the North Celestial Pole for the Northern hemisphere. In the Northern Hemisphere Polaris is very close to the North Celestial Pole and provides an adequate position for observing. The basic aim of Polar Alignment is to align the telescope mounts Right Ascension (RA) axis to Polaris. The simplest method of polar alignment is simply to aim the RA axis at Polaris. In some cases, star trail photographs are desired. In that case, no tracking should be done.

Tripods, Mounts, and Tracking

The photographic tripod or alt-azimuth mount is the simplest mount. The alt-azimuth mount has two axes of movement: a horizontal axis and a vertical axis. To point the telescope at an object, move it along the horizon (azimuth axis) to the object's horizontal position, and then tilt the telescope, along the altitude axis, to the object's vertical position. It does not align and track stars. This type of mount can be used for binoculars, spotting scopes, cameras, and low power telescopes.

The equatorial mount has two polar aligned axes of rotation: right ascension and declination. However, instead of being oriented up and down, it is tilted at the same angle as the Earth's axis of rotation. This type of mount can track stars manually or automatically when connected to a clock mechanism.

These types of mounts are available in manual or computerized Go-to operation.

Vibration pads can also be used to minimize tripod vibration.

Camera Vibration and Exposure Time

A very important aspect of astrophotography for either camera or telescopes is that the mount MUST be STURDY. Even a micro amount of movement or shake can result in burred photos.

A camera can be mounted on a tracking mount, a non-tracking tripod, or on a telescope, which is mounted to a tracking or non-tracking mount. The earth rotates on its axis at about 1/2-degree per minute. The maximum exposure time will depend upon the focal length of the lens or telescope, the ISO speed of the camera, the pixel size of the sensor, and the declination. Generally, the maximum exposure time is 5 minutes.

A remote shutter release control should always be used to avoid shaking the camera while pressing the shutter release button on the camera. A remote shutter release intervalometer is an inexpensive device that can be set to expose one or multiple exposures over a variable period of time.

DSLRs allow light to enter through the lens. Then, the light is reflected by a mirror to the viewfinder. When, the shutter button is pressed, the mirror flips so that the light will be projected onto a digital sensor or film. When, the mirror moves, the camera will shake, which can result in burred images for long exposure times. Many DSLR cameras have a mirror lockup feature to prevent this. They also have an exposure delay function to prevent camera vibrations during long exposures. The mirror up function stops the mirror from moving when capturing the image to prevent camera vibration. (2 shutter releases for Nikon cameras). The Exposure Delay function waits 1, 2, or 3 seconds after the mirror is raised to capture the image to prevent camera vibration.

ISO, formerly known as ASA, expresses the speed or sensitivity of film or a digital sensor to light. The higher the ISO, the more sensitive the image sensor and therefore the possibility to take pictures in low-light situations. Increasing ISO will decrease exposure time. However, increasing ISO also increases sensor noise or grain in film. For finer detail, use an ISO of 100 or 200. In some cases, this can make the exposure time too long. To accommodate circumstances, increase the ISO. Some digital DSLR cameras today offer ISO speeds as high as 28,000.

Several photos were tested on 5/9/17 at 1/500 second and mirror up made a difference.

Correctors and Reducers

Some telescopes have spherical aberrations and some telescopes have chromatic aberrations. Field flatters, reducers, and chromatic corrects can be used to correct these problems.

Generally F6 doesn't require coma corrector while F4 requires a coma corrector

Software Guiding

There are a number of software programs that will time and guide a mount and camera. PHD Guiding is popular and available for mac and pc. Computerized mounts will also guide a camera and telescope.

Planning Equipment with Software

Planning what is going to be photographed and with what equipment can be done easily on a computer using the program "Stallarium." Stallarium can be downloaded for free from http://www.stellarium.org

The formula for determining the size of the lunar images on a sensor is found by dividing the focal length (in millimeters) by 109 to arrive at the lunar diameter (in millimeters) on the sensor.

For a DX Format, 23.5 mm x 15.6 mm, 1700 mm = full frame height

For a DX Format, 23.5 mm x 15.6 mm, 850 mm = half frame height

For an FX Format, 35.9 mm x 24 mm, 2616 mm = full frame height

For an FX Format, 35.9 mm x 24 mm, 1308 mm = half frame height

The Stallarium simulation below represents an ideal photograph of the moon using the Celestron C6 (150 mm x 1500 mm) SCT telescope and the Nikon D7100 camera. The camera frame is represented by the red rectangle surrounding the moon.

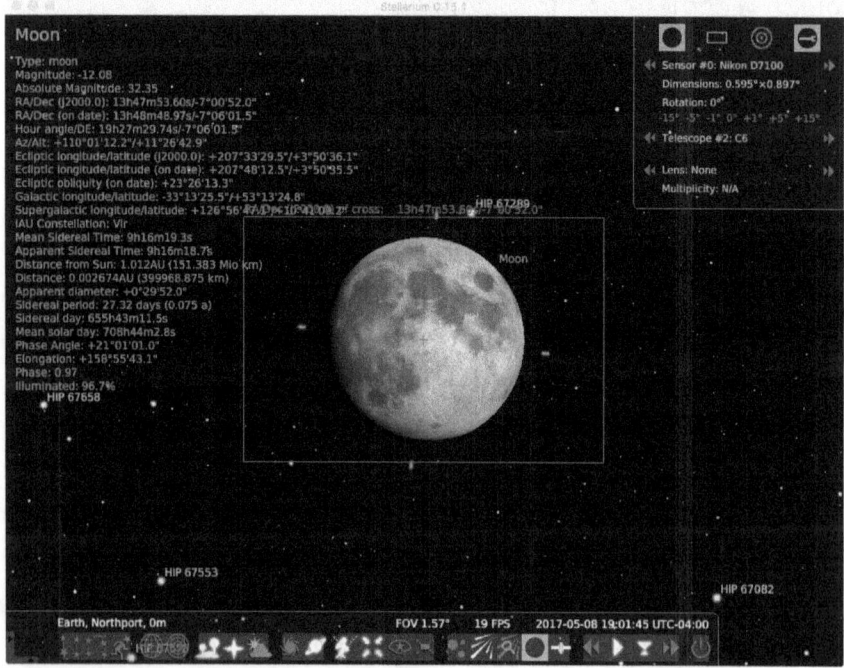

Telescope and Camera Configurations

The diagram below shows a camera connected to the prime focus using a prime focus and T adapter if it is a DSLR. The T adapter is specific for each camera.

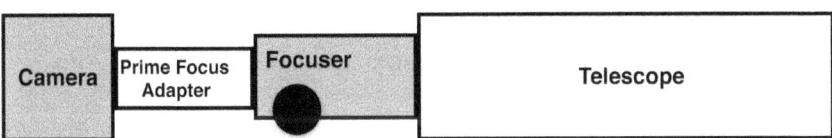

The diagram below shows a camera connected to the eyepiece adapter. If the telescope has a 2" focuser, the prime focus connection will create a better image than the 1.25" eyepiece adapter. Although, using an eyepiece adapter may allow for an eyepiece to be inserted to provide additional magnification.

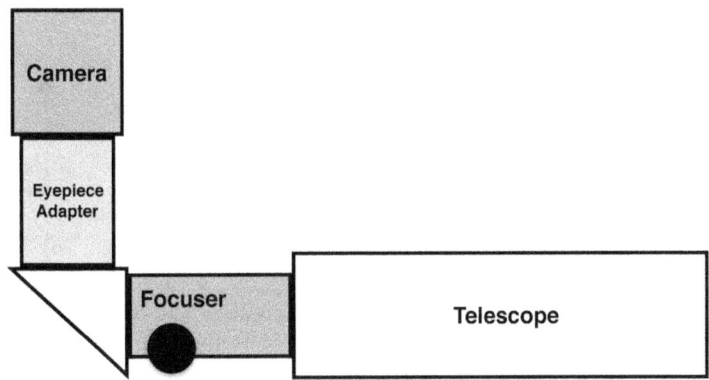

The next problem to solve is what optics to use? There are many different cameras, telescopes, eyepieces, barlow lenses, teleconverters, etc. For wide field astrophotography, many astronomers use only a camera and wide angle lens.

Objects, Cameras, Lenses, and Telescope Simulations

Note: There is no difference in focal length or magnification between crop sensors and full frame sensors. ONLY THE FIELD OF VIEW (FOV) changes. In the simulations below, the field of view is changed by the Stallarium software to provide a standardized view of the object.

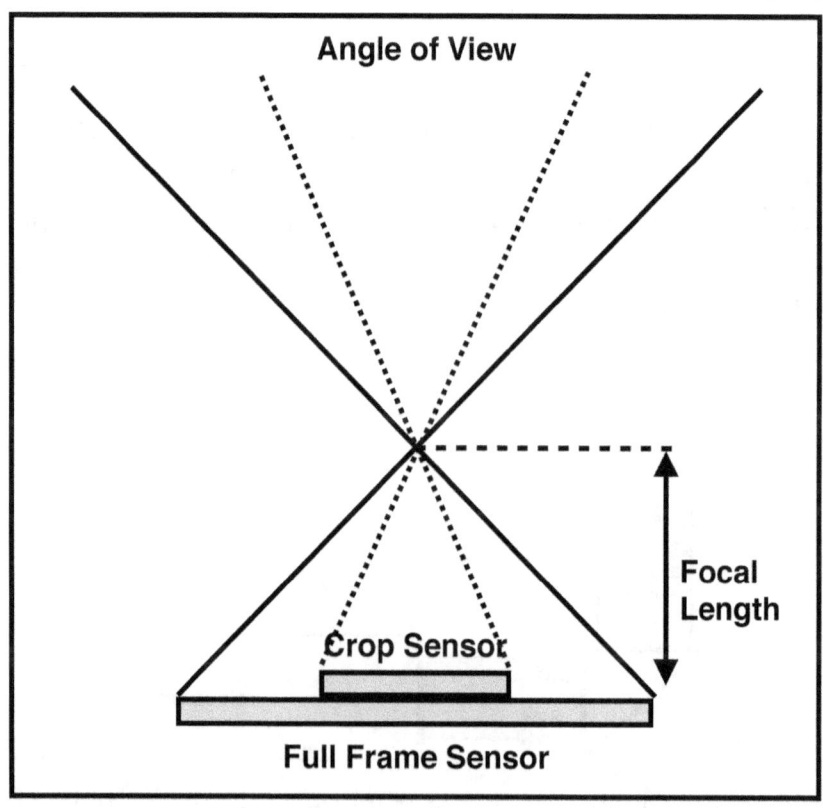

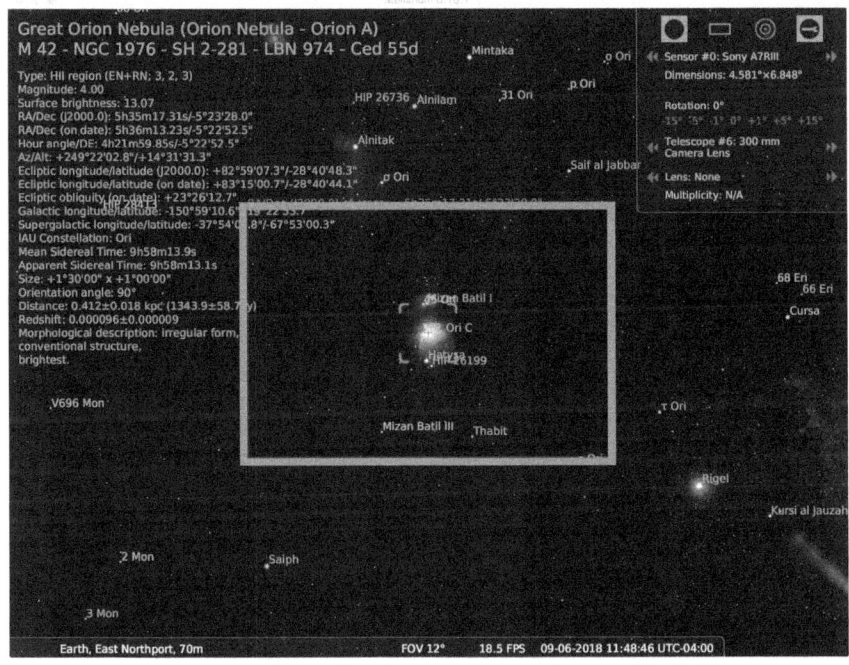

Orion Nebula (M42) with 300 mm lens, Full Frame
Camera, FOV 12 degrees

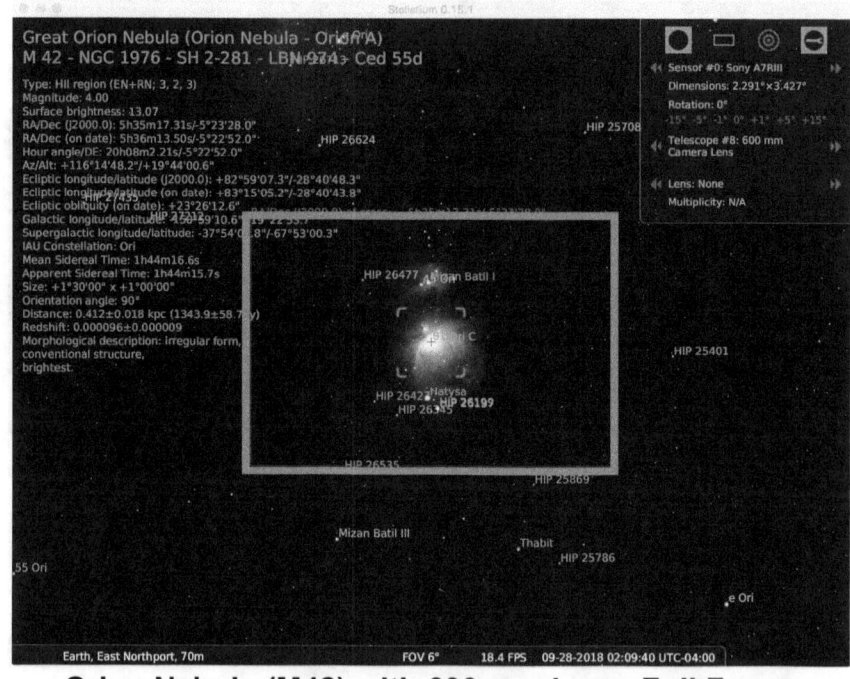

Orion Nebula (M42) with 600 mm Lens, Full Frame Camera, FOV 6 degrees

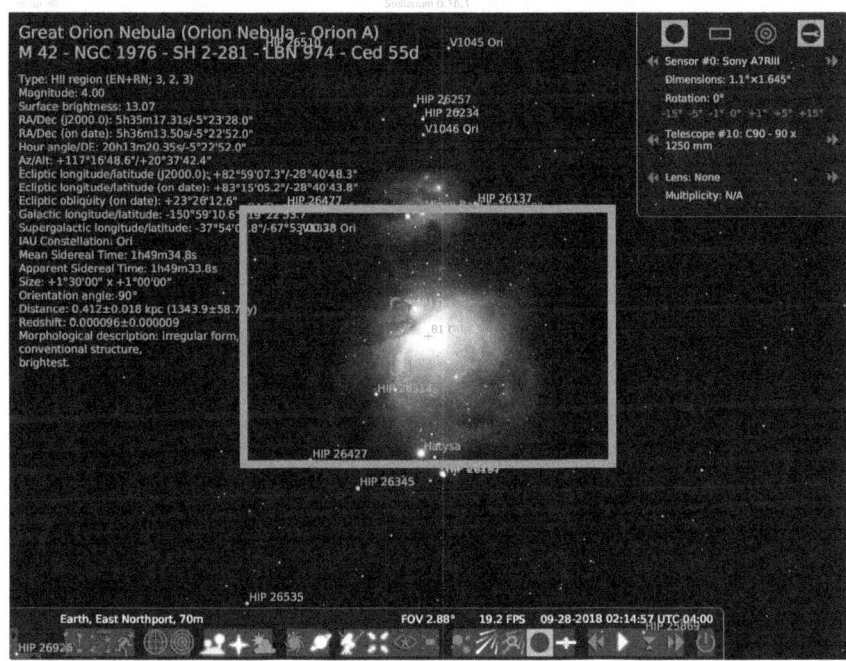

Orion Nebula (M42) with C90 1250 mm Telescope, Full Frame Camera, FOV 2.88 degrees

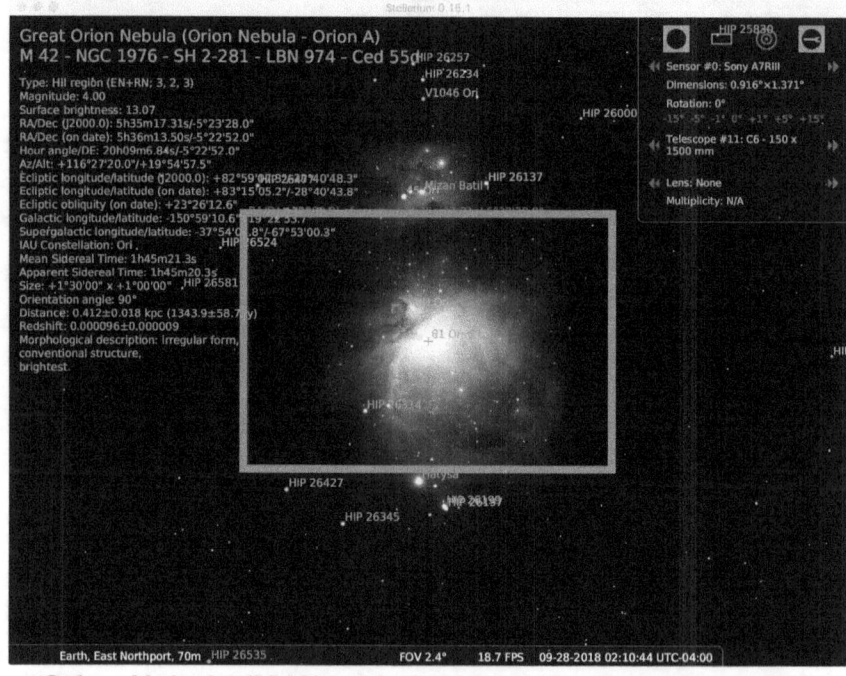

Orion Nebula (M42) with C6 1500 mm Telescope, Full Frame Camera, FOV 2.4 degrees

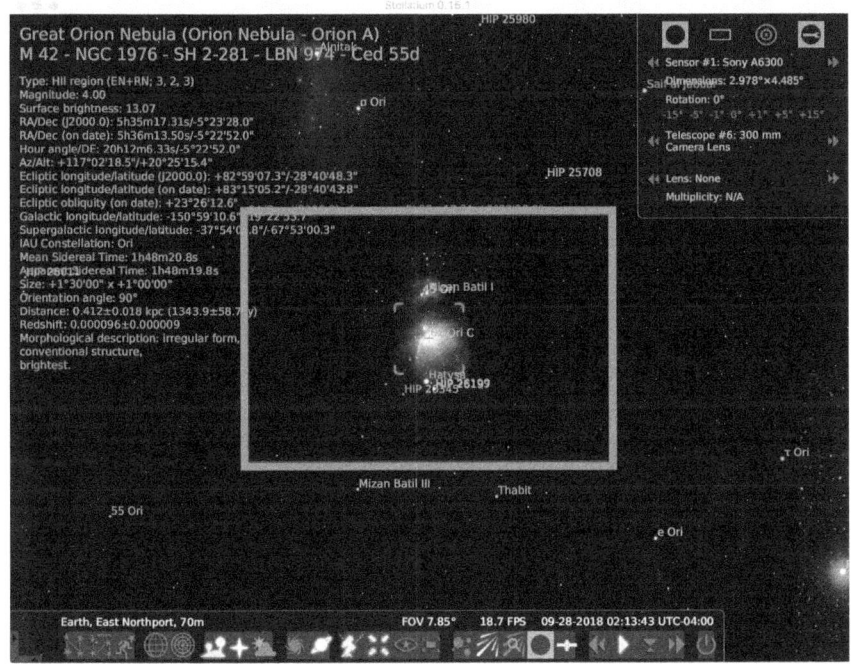

Orion Nebula (M42) with 300 mm lens, APS-C Camera, FOV 7.85 degrees

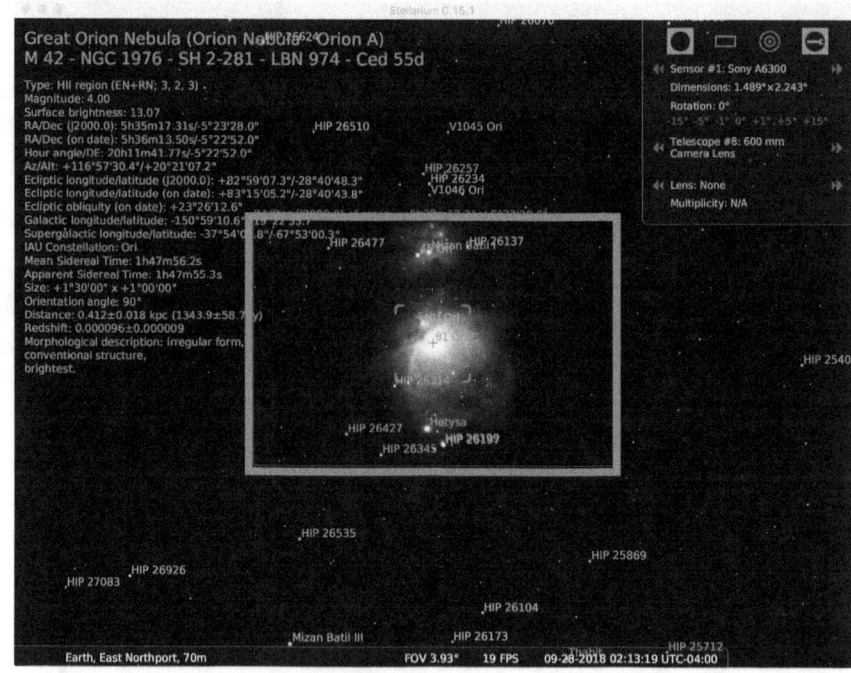

Orion Nebula (M42) with 600 mm lens, APS-C Camera, FOV 3.93 degrees

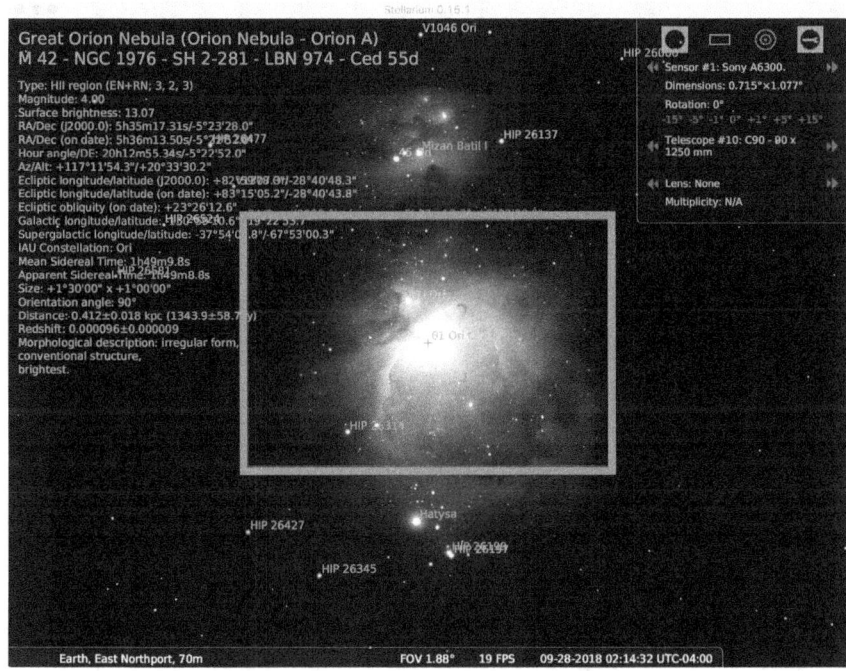

Orion Nebula (M42) with C90 1250 mm Telescope, APS-C Camera, FOV 1.88 degrees

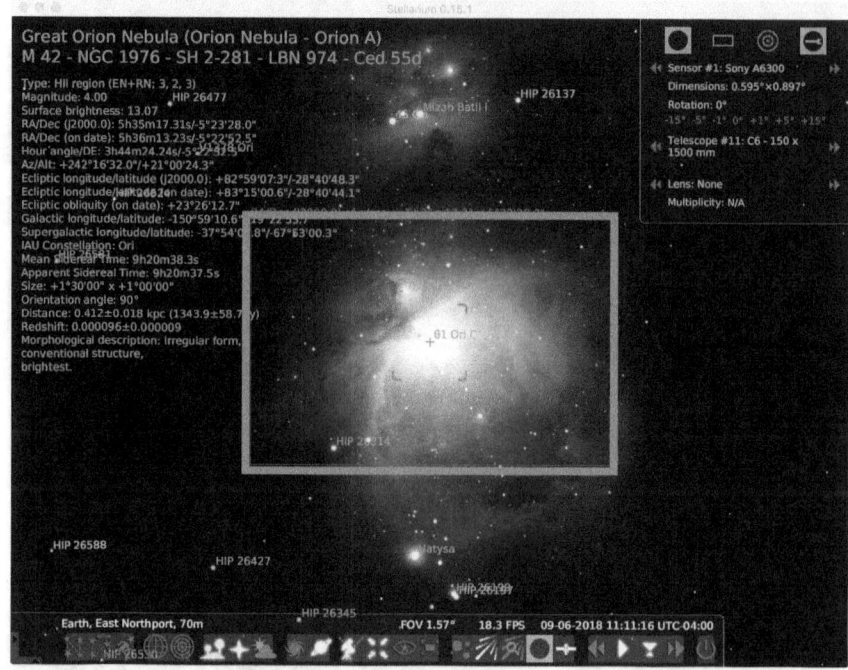

Orion Nebula (M42) with C6 1500 mm Telescope, APS-C
Camera, FOV 1.57 degrees

144

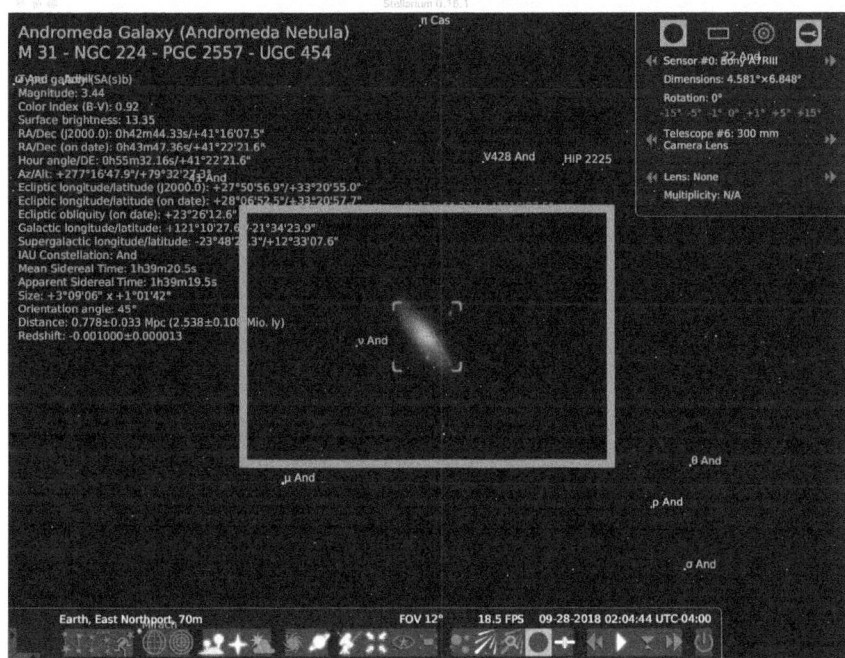

Andromeda Galaxy (M31) with 300 mm lens, Full Frame Camera, FOV 12 degrees

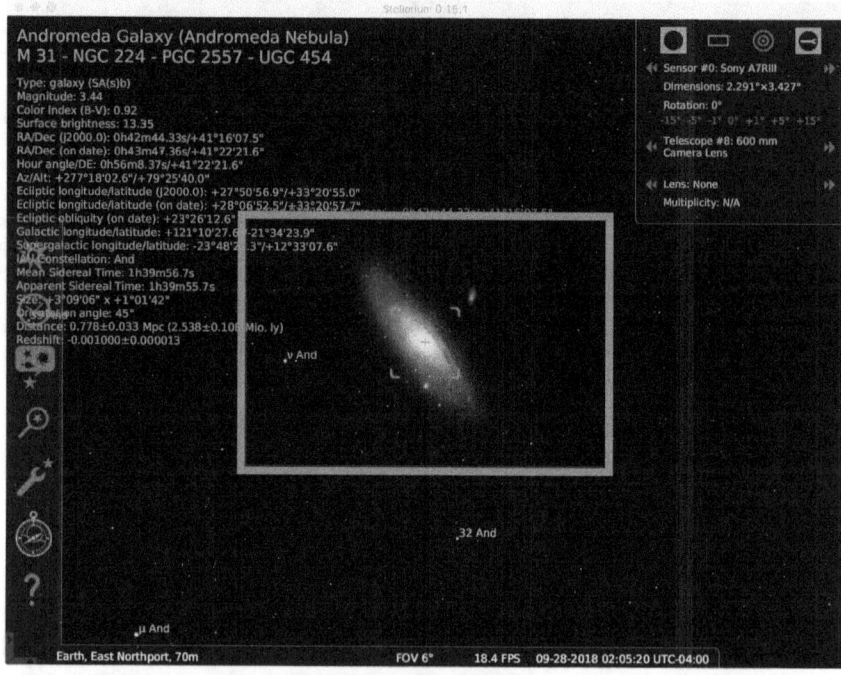

Andromeda Galaxy (Andromeda Nebula)
M 31 - NGC 224 - PGC 2557 - UGC 454

Type: galaxy (SA(s)b)
Magnitude: 3.44
Color Index (B-V): 0.92
Surface brightness: 13.35
RA/Dec (J2000.0): 0h42m44.33s/+41°16'07.5"
RA/Dec (on date): 0h43m47.36s/+41°22'21.6"
Hour angle/DE: 0h56m8.37s/+41°22'21.6"
Az/Alt: +277°18'02.6"/+79°25'40.0"
Ecliptic longitude/latitude (J2000.0): +27°50'56.9"/+33°20'55.0"
Ecliptic longitude/latitude (on date): +28°06'52.5"/+33°20'57.7"
Ecliptic obliquity (on date): +23°26'12.6"
Galactic longitude/latitude: +121°10'27.6"/-21°34'23.9"
Supergalactic longitude/latitude: -23°48'2 .3"/+12°33'07.6"
IAU Constellation: And
Mean Sidereal Time: 1h39m56.7s
Apparent Sidereal Time: 1h39m55.7s
Size: +3°09'06" x +1°01'42"
Orientation angle: 45°
Distance: 0.778±0.033 Mpc (2.538±0.10 Mio. ly)
Redshift: -0.001000±0.000013

Sensor #0: Sony A7RIII
Dimensions: 2.291"x3.427"
Rotation: 0°
-15° -5° -1° 0° +1° +5° +15°
Telescope #8: 600 mm
Camera Lens
Lens: None
Multiplicity: N/A

ν And

32 And

μ And

Earth, East Northport, 70m   FOV 6°   18.4 FPS   09-28-2018 02:05:20 UTC-04:00

Andromeda Galaxy (M31) with 600 mm lens, Full Frame Camera, FOV 6 degrees

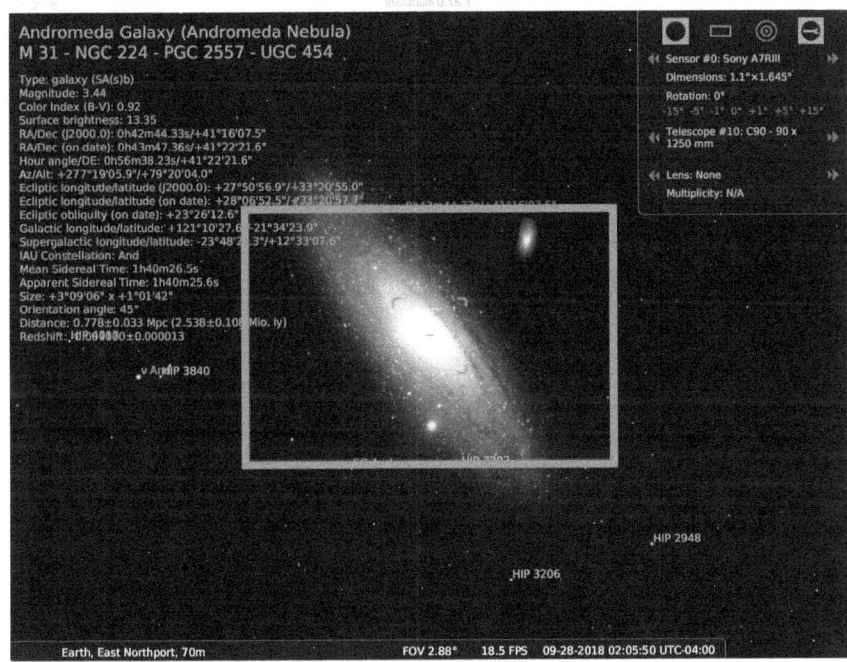

Andromeda Galaxy (M31) with C90 1250 mm Telescope, Full Frame Camera, FOV 2.88 degrees

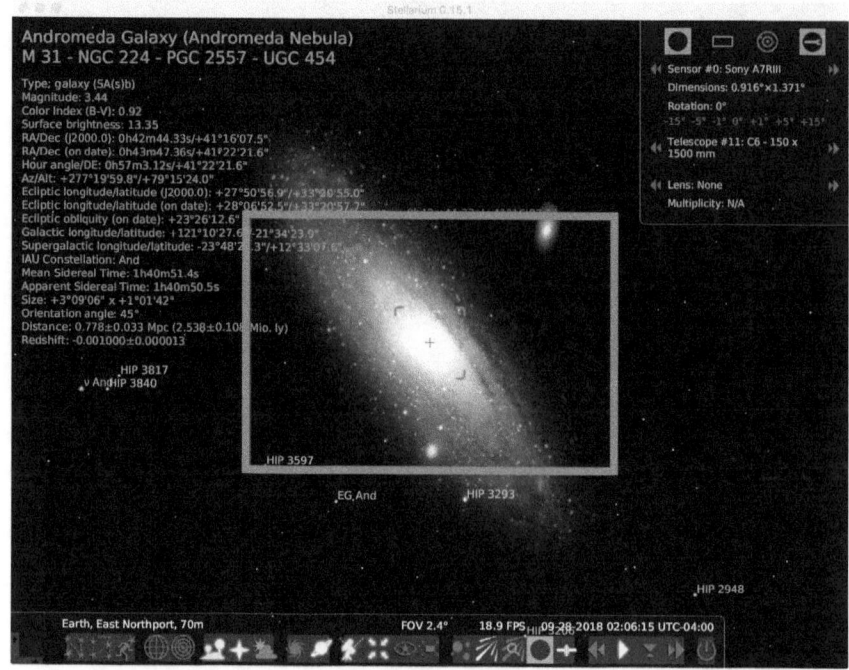

Andromeda Galaxy (M31) with C6 1500 mm Telescope, Full Frame Camera, FOV 2.4 degrees

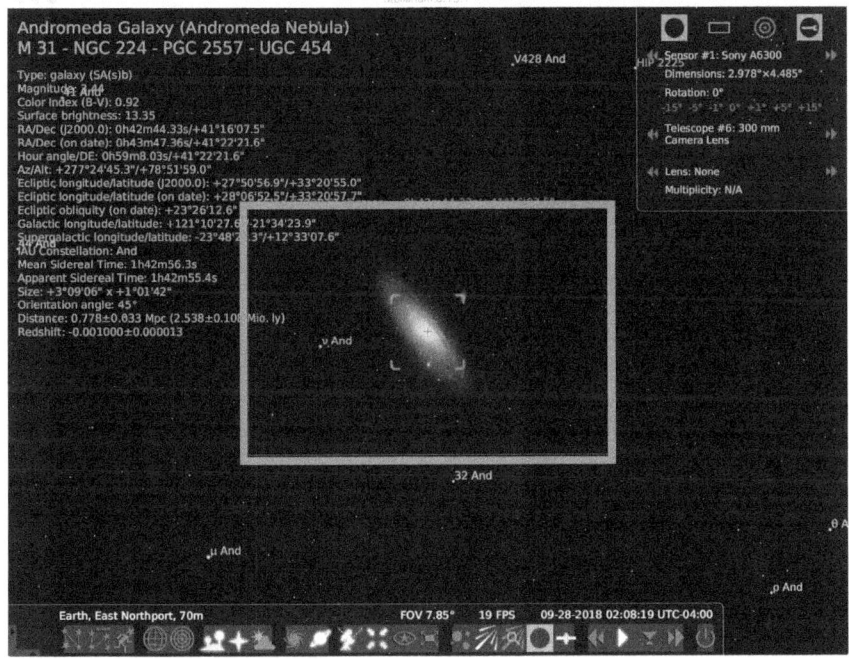

Andromeda Galaxy (M31) with 300 mm lens, APS-C Camera, FOV 7.85 degrees

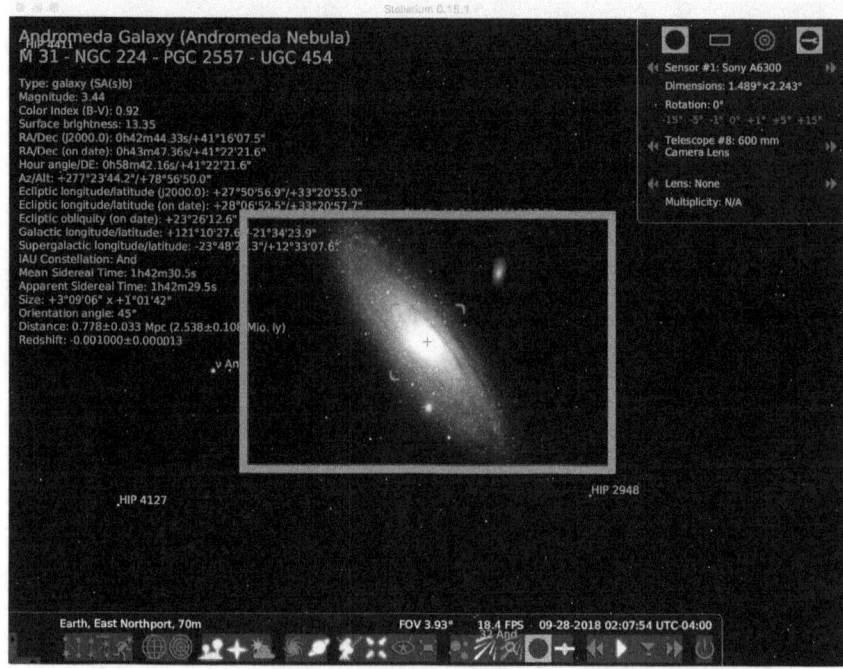

**Andromeda Galaxy (M31) with 600 mm lens, APS-C
Camera, FOV 3.93 degrees**

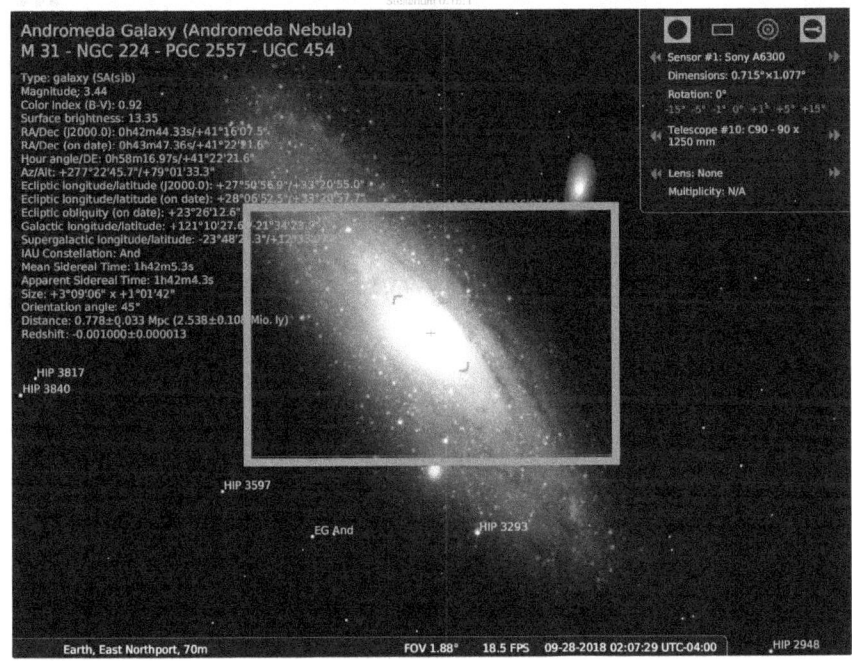

Andromeda Galaxy (M31) with C90 1250 mm Telescope, APS-C Camera, FOV 1.88 degrees

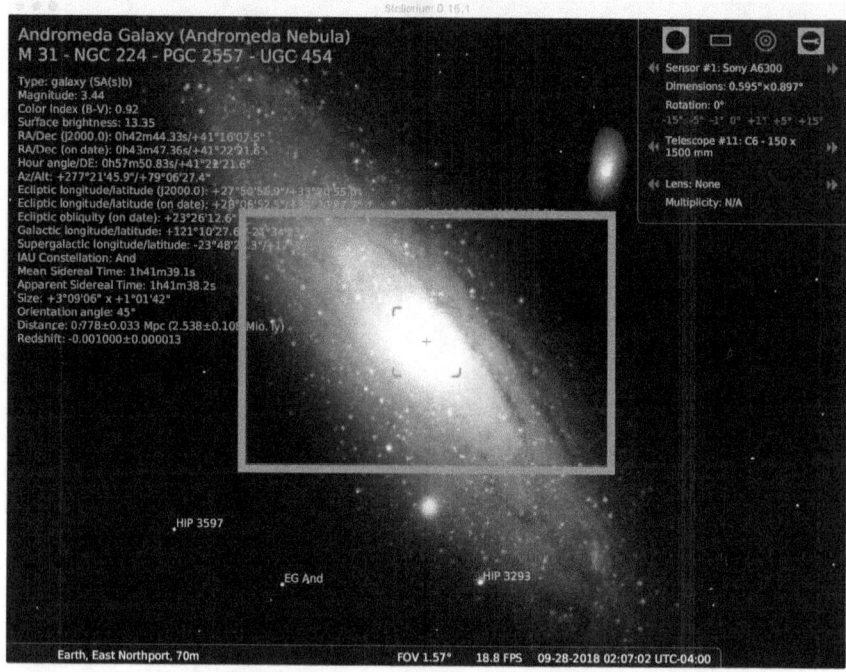

Andromeda Galaxy (Andromeda Nebula)
M 31 - NGC 224 - PGC 2557 - UGC 454

Type: galaxy (SA(s)b)
Magnitude: 3.44
Color Index (B-V): 0.92
Surface brightness: 13.35
RA/Dec (J2000.0): 0h42m44.33s/+41°16'07.5"
RA/Dec (on date): 0h43m47.36s/+41°22'21.6"
Hour angle/DE: 0h57m50.83s/+41°22'21.6"
Az/Alt: +277°21'45.9"/+79°06'27.4"
Ecliptic longitude/latitude (J2000.0): +27°50'56.9"/+33°20'55.3"
Ecliptic longitude/latitude (on date): +28°06'52.5"/+33°27'24.9"
Ecliptic obliquity (on date): +23°26'12.6"
Galactic longitude/latitude: +121°10'27.6"/-21°34'12"
Supergalactic longitude/latitude: -23°48'2.3"/+1°1"
IAU Constellation: And
Mean Sidereal Time: 1h41m39.1s
Apparent Sidereal Time: 1h41m38.2s
Size: +3°09'06" x +1°01'42"
Orientation angle: 45°
Distance: 0.778±0.033 Mpc (2.538±0.108 Mio. ly)
Redshift: -0.001000±0.000013

Sensor #1: Sony A6300
Dimensions: 0.595°×0.897°
Rotation: 0°
-15° -5° -1° 0° +1° +5° +15°

Telescope #11: C6 - 150 x
1500 mm

Lens: None
Multiplicity: N/A

HIP 3597

EG And

HIP 3293

Earth, East Northport, 70m   FOV 1.57°   18.8 FPS   09-28-2018 02:07:02 UTC-04:00

Andromeda Galaxy (M31) with C6 1500 mm Telescope, APS-C Camera, FOV 1.57 degrees

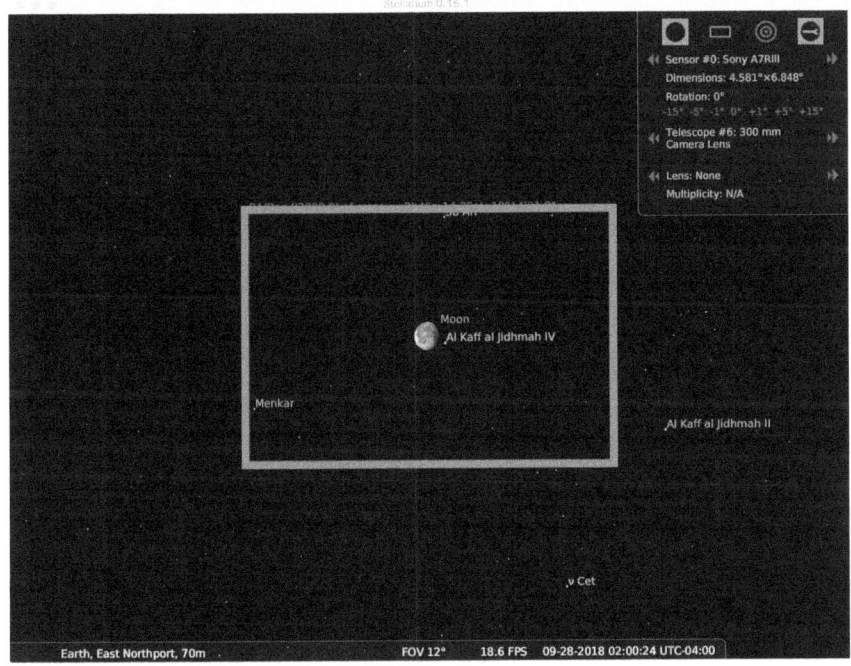

Moon with 300 mm lens, Full Frame Camera, FOV 12 degrees

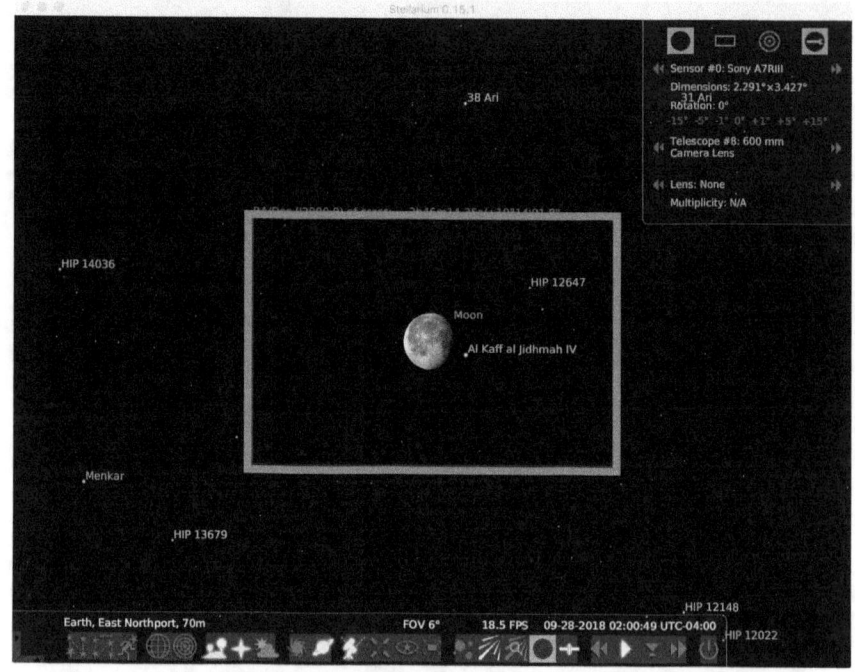

Moon with 600 mm lens, Full Frame Camera, FOV 6 degrees

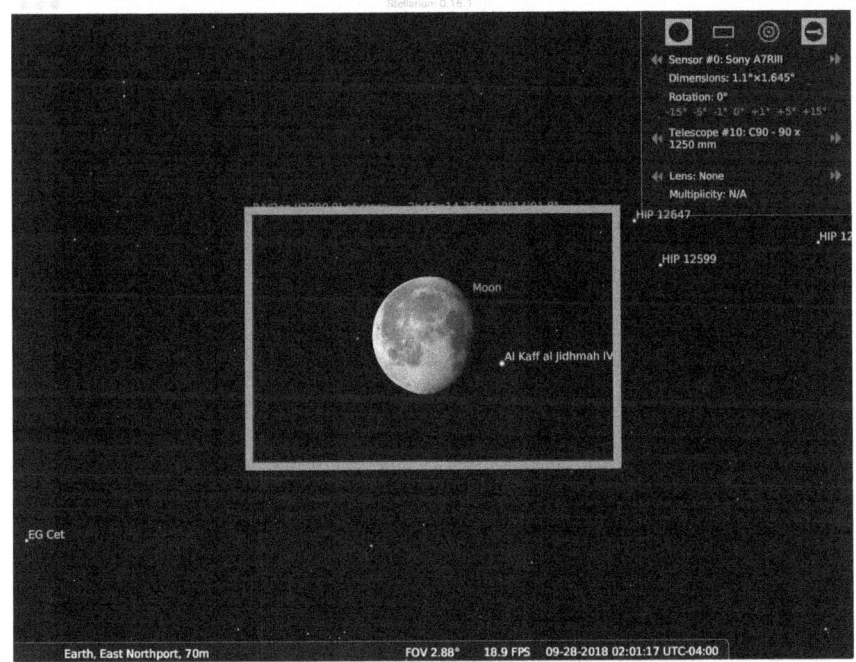

Moon with C90 1250 mm Telescope, Full Frame Camera, FOV 2.88 degrees

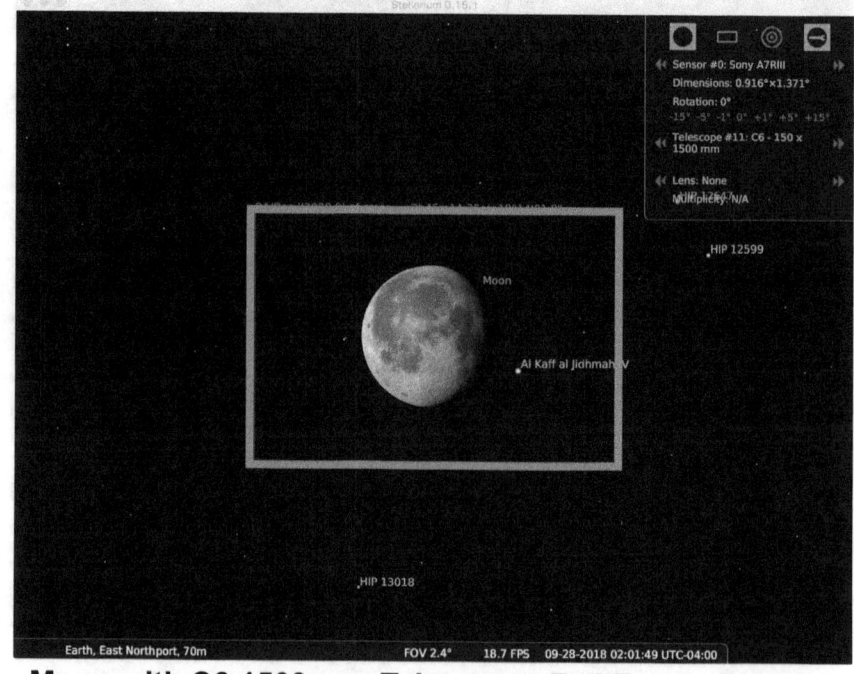

Moon with C6 1500 mm Telescope, Full Frame Camera, FOV 2.4 degrees

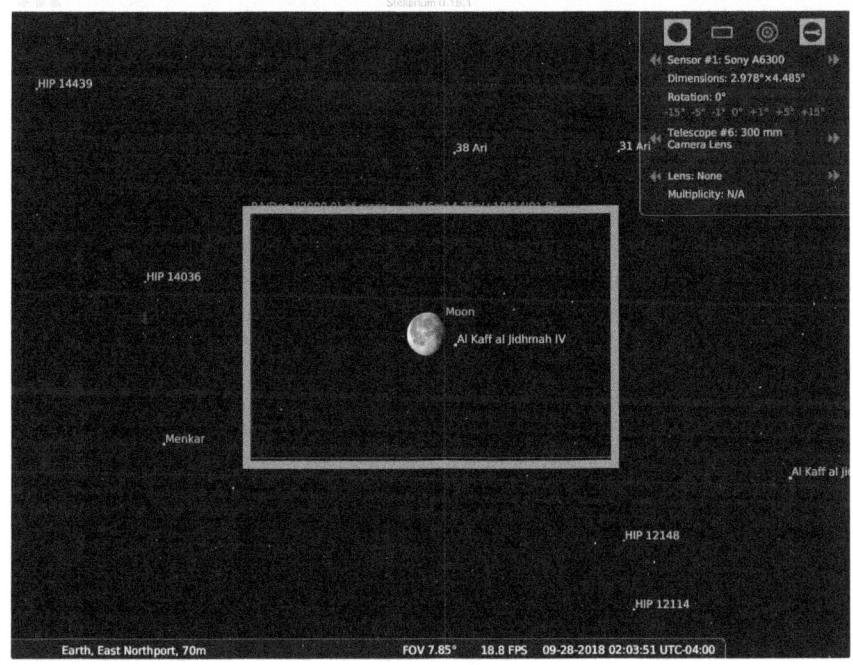

Moon with 300 mm lens, APS-C Camera, FOV 7.85 degrees

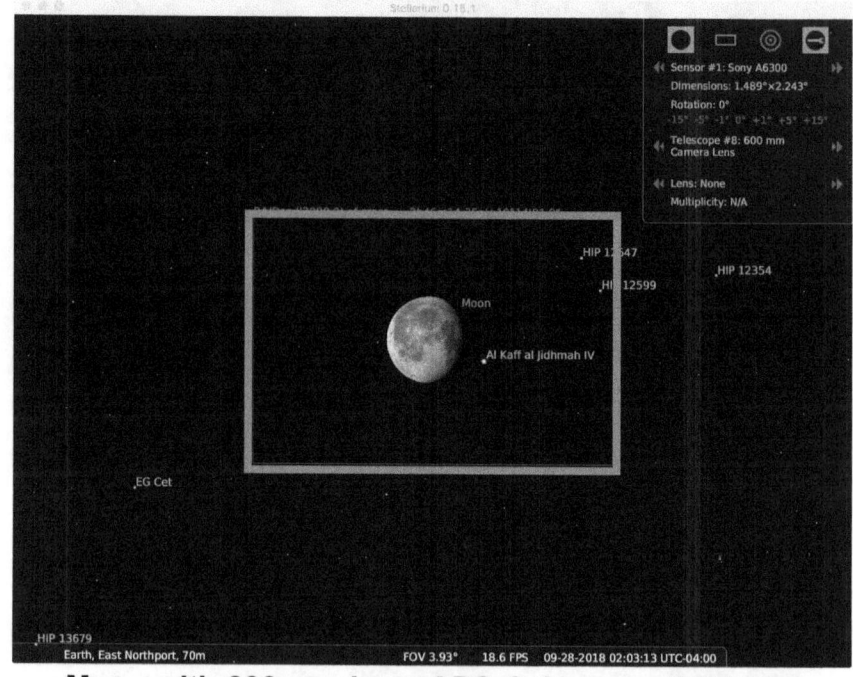

Moon with 600 mm lens, APS-C Camera, FOV 3.93 degrees

Moon with C90 1250 mm Telescope, APS-C Camera, FOV 1.88 degrees

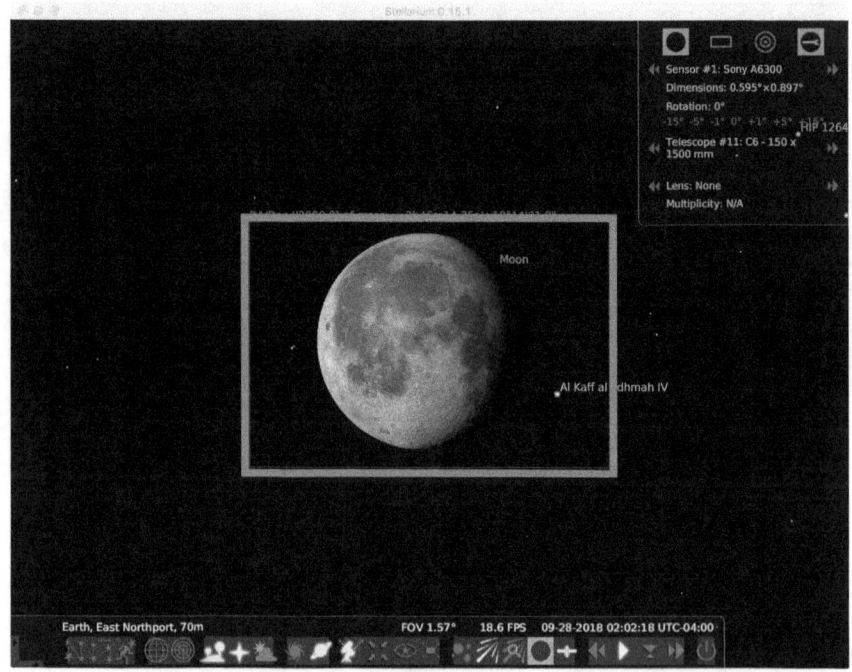

Moon with C6 1500 mm Telescope, APS-C Camera, FOV 1.57 degrees

Nikon Mirror Up Mode

1. Select mirror up mode on the mode dial.

2. Frame the picture, focus, then press the shutter release button all the way down to raise the mirror.

3. Press the shutter release button all the way a second time to take the picture. Use the remote shutter release to prevent camera shake.

Note: if a camera doesn't have a mirror up mode, the "Exposure Delay Mode" is an alternative. Ignore this for mirrorless cameras.

Post-processing in Photoshop

Photoshop can be a very useful tool is adjusting brightness, contrast, and sharpness of a photograph. It can also be used to stack multiple photographs together.

Photographing the Moon

Moon photography is often a recommended place to start because it is "easy." Well, it may not be all that easy. It could be a matter of trial and error. By comparison to deep space objects, the moon gives off a lot of light. Therefore, tracking is not required. This makes it easier than DSO photography. Tracking adds a great deal of difficulty to the equation.

The moon moves 1/2 degree, in about 2 minutes. As a general rule, the moon can be photographed for about 2 seconds before the moon's movement becomes a problem. A camera and telephoto lens may be the best equipment to start with. Later, a telescope can be used with adapters, etc.

Teleconverters are often incompatible with certain lenses and cameras. Check the manufacturer's compatibility list before purchasing.

The rule of thumb for a tripod / mount is that it should have a payload capacity of twice the weight of the payload (telescope). Equatorial go-to tracking mounts for telescopes that weigh 10 pounds can weigh in excess of 50 pounds. Camera go-to tracking mounts weigh far less.

The Stallarium simulation below represents an ideal photograph of the moon using the Celestron C6 (150 mm x 1500 mm) SCT telescope and the Nikon D7100 camera. The camera frame is represented by the red rectangle surrounding the moon. The magnification appears to be ideal.

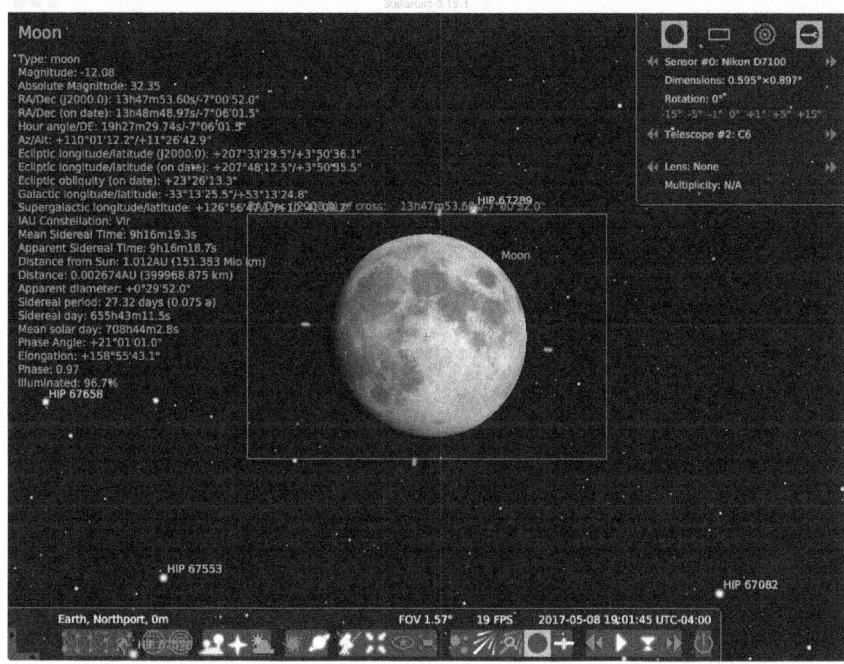

Moon

Type: moon
Magnitude: -12.08
Absolute Magnitude: 32.35
RA/Dec (J2000.0): 13h47m53.60s/-7°00'52.0"
RA/Dec (on date): 13h48m48.97s/-7°06'01.5"
Hour angle/DE: 19h27m29.74s/-7°06'01.5"
Az/Alt: +110°01'12.2"/+11°26'42.9"
Ecliptic longitude/latitude (J2000.0): +207°33'29.5"/+3°50'36.1"
Ecliptic longitude/latitude (on date): +207°48'12.5"/+3°50'55.5"
Ecliptic obliquity (on date): +23°26'13.3"
Galactic longitude/latitude: -33°13'25.5"/+53°13'24.8"
Supergalactic longitude/latitude: +126°56'47.3"/+10°01'08.2"
IAU Constellation: Vir
Mean Sidereal Time: 9h16m19.3s
Apparent Sidereal Time: 9h16m18.7s
Distance from Sun: 1.012AU (151.383 Mio km)
Distance: 0.002674AU (399968.875 km)
Apparent diameter: +0°29'52.0"
Sidereal period: 27.32 days (0.075 a)
Sidereal day: 655h43m11.5s
Mean solar day: 708h44m2.8s
Phase Angle: +21°01'01.0"
Elongation: +158°55'43.1"
Phase: 0.97
Illuminated: 96.7%

HIP 67658
HIP 67289
HIP 67553
HIP 67082

Sensor #0: Nikon D7100
Dimensions: 0.595°×0.897°
Rotation: 0°
-15° -5° -1° 0° +1° +5° +15°
Telescope #2: C6
Lens: None
Multiplicity: N/A

Moon

Earth, Northport, 0m   FOV 1.57°   19 FPS   2017-05-08 19:01:45 UTC-04:00

The moon can even be photographed on a cloudy night with an iPhone. The image below was taken with an iPhone 7 on a cloudy night and processed in Photoshop to increase brightness, contrast, and sharpness. The unsharp filter mask was used and set at 150% (Filter > Sharpen > Unsharp Mask > Amount = 150%).The fence, lower portion of the house, garage, and shed were cropped from the bottom of the photograph. Metadata = f/1.8, 1/15, ISO 800, 3024 x 4032.

Cameras, Telescopes, and Lens

For astrophotography, generally, a lens of 8-24 mm focal length with f/2.8 aperture or greater for wide field photographs is recommended.

For planets, the moon, and deep sky objects, a telescope with a focal length of at least 600 mm is recommended.

For wide field images, a full frame sensor camera is recommended.

14 mm lens on a Full Frame sensor camera will produce an image of 114 degrees wide, which is wider than what the human eye can see and subject to perspective distortion.

20 mm lens on a Full Frame sensor camera will produce an image that is 94 degrees wide, which is about what the human eye sees.

24 mm lens on a Full Frame sensor camera will produce an image of 84 degrees wide, which is less than what a human eye can see and is used on most standard zoom lenses.

Configuration 1 - DSLR with Telephoto Lens

The configuration below is only one of the many different configurations that are possible. Choose what you want to configure and how much you want to spend. There are an almost infinite number of options available. The resource section adds more possibilities. A teleconverter would not be compatible with the 300 mm lens.

Nikon D7100 DSLR Camera (Crop Sensor DSLR Camera)
24.1 MP DX-Format CMOS Sensor, $700

Nikon AF-S VR Zoom-Nikkor 70-300 mm f/4.5-5.6G IF-ED Camera Lens, $500 (or Non ED lens for $200)

Olivon TR197-16 Tripod with Head (Tall Tripod)
Total Weight is 15 pounds, $250

Celestron Vibration Suppression Pads, ITEM # 93503, $50

Vello ShutterBoss II Timer Remote Switch for Nikon with DC2 Connection, $50

Total $1,550 ($1,250 with non ED Lens)

Configuration 2 - 6 inch Telescope and Prime Focus DSLR

The configuration below is only one of the many different configurations that are possible. Choose what you want to configure and how much you want to spend. There are an almost infinite number of options available. The resource section adds more possibilities.

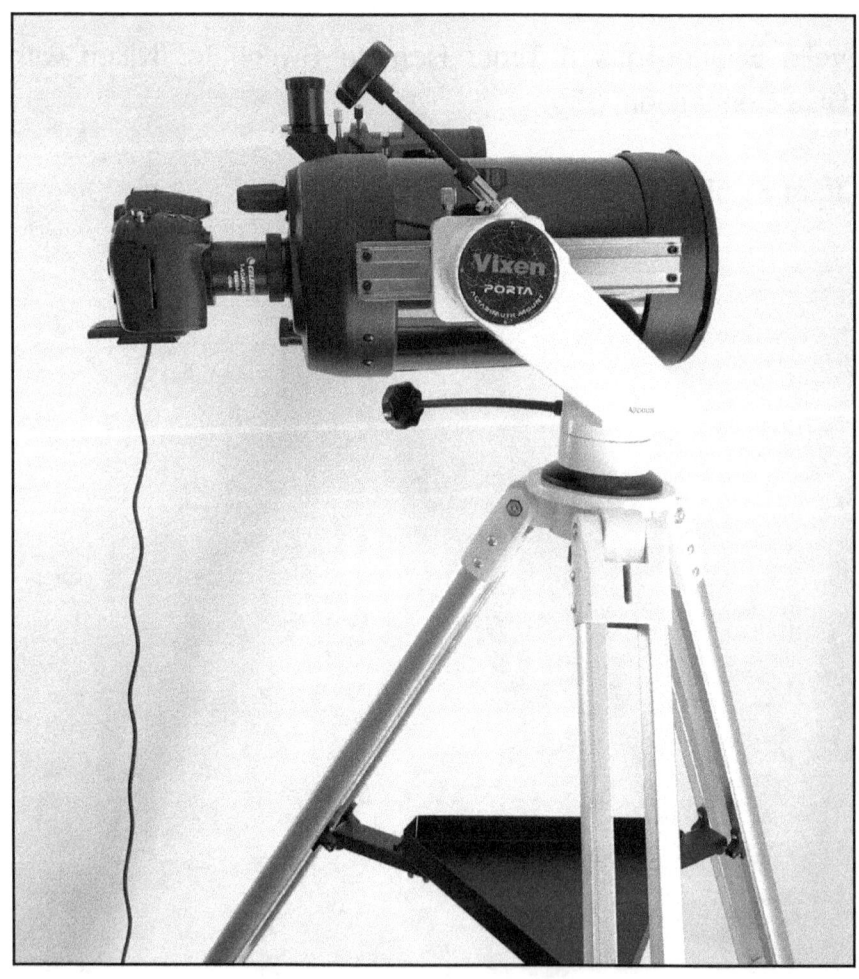

Celestron C6-A-XLT CG-5 6" f/10 Schmidt-Cassegrain Telescope (OTA), $600
or substitute the C90 MCT Spotting Scope for $200

Celestron 94009 Lens Shade for C6 and C8 Tubes (Black), Item 94009, $25 (not for C90)

Celestron T-Adapter for Schmidt Cassegrain Telescopes (for C6 SCT), ITEM # 93633-A, $25 (not require for C90)

Vello Lens Mount Adapter - T Mount Lens to Nikon F Mount Camera, LANFT, $14

Vixen Porta II Mount Tall, Total Weight is 13.9 pounds, $400

Celestron Vibration Suppression Pads, ITEM # 93503, $50

Nikon D7100 DSLR Camera (Crop Sensor DSLR Camera), $700

Vello ShutterBoss II Timer Remote Switch for Nikon with DC2 Connection, $50

Total $1,864

Photographing the Moon with Configuration 1 - DSLR with Telephoto Lens

1. Attach the telephoto lens to the camera body.

2. Extend the tripod legs, insert the vibration pads under, and attach the camera.

3. Insert the Vello remote switch into the camera.

4. Set the camera to mirror lockup and live view with magnification.
   a. Select mirror up mode on the mode dial.
   b. Frame the picture, focus, then press the shutter release button all the way down to raise the mirror.
   c. Press the shutter release button all the way a second time to take the picture. Use remote shutter release to prevent camera shake.

5. Set the camera to manual and the camera lens to manual focus.

6. Set the ISO to 200.

7. Set the shutter speed to 1/125 seconds or higher. This may need to be adjusted faster or slower.

8. Set the telephoto lens to maximum focal length (300 mm or 600 mm)

9. Aim the camera at the moon.

10. Focus the camera using Live View with magnification.

11. Press the Vello remote switch button and take several photographs, maybe 20 photographs.

12. View the photographs taken. Adjust ISO and shutter speeds if necessary and repeat. Auto ISO is an alternative.

Photographing the Moon with Configuration 2 - Telescope and DSLR

1. If the diagonal is attached to the telescope, remove it.

2. Attach the Vello Lens Mount Adapter to the Celestron T-Adapter

3. Attach one end of the T-Adapter to the camera.

4. Screw the other end of the T-Adapter into the telescope.

5. Extend the legs of the tripod, attach the mount to the tripod, and insert the vibration pads under.

6. Attach the telescope to the mount.

7. Insert the Vello remote switch into the camera.

8. Set the camera to mirror lockup and live view with magnification.
   a. Select mirror up mode on the mode dial.
   b. Frame the picture, focus, then press the shutter release button all the way down to raise the mirror.
   c. Press the shutter release button all the way a second time to take the picture. Use remote shutter release to prevent camera shake.

9. Set the camera to manual and the camera lens to manual focus.

10. Set the ISO to 200. Auto ISO is an alternative.

11. Set the shutter speed to 1/125 seconds or higher. This may need to be adjusted faster or slower.

12. Aim the telescope at the moon.

13. Focus the telescope using Live View on the camera.

14. Press the Vello remote switch button and take several photographs, maybe 20 photographs.

15. If a digital DSLR is used, view the photographs taken. Adjust ISO and shutter speeds if necessary and repeat. Auto ISO is an alternative.

Astrophotography Without Tracking

Is a camera an accessory for a telescope or is a telescope a telephoto lens for a camera? For the following discussion, let the telescope be a telephoto lens for a camera and in this case, a DSLR camera.

Tracking is delicate, tricky, and sometimes a disaster. Imagine the accuracy that is required when photographing over a period of hours. This discussion avoids the use of tracking for most amateur astrophotography. There are many astro-photographs posted on the internet showing excellent astrophotography, including deep space objects (DSO), without tracking.

The light coming from the moon is bright and often requires only a 1/125 second exposure at ISO 200. The light coming from stars in the night sky is very dim and used to require extremely long exposure times, often hours long. However, todays cameras have digital sensors that are far more sensitive than film. Kodachrome had ISO speed from 25 to 200 in the 1970s. The Nikon D7100 digital camera, has ISO speeds up to 25,600. Higher ISO speeds mean shorter exposures. Higher ISO speeds also mean higher sensor noise. Many astrophotographers get excellent results using an ISO speed of 6,400.

Taking multiple short exposure images and stacking them together using a software program will produce an excellent photograph. There are many software programs available for mac and pc. Stacking increases the signal-to-noise ratio and increases the dynamic range. There are a number of theories as to why this happens. Whatever the reason, stacking produces better photographs. How many images to stack? Twenty to one hundred photographs seems like a good number. Each photograph should have adequate exposure to stand on its own. Experiment on your own and see what you come up with.

According to Erwin Matys, Karoline Mrazek (project nightflight), the formula to calculate the MAXIMUM exposure time is: $t = (27000 \times \Delta) / (f \times \cos \delta)$ in the article: "DSLR Astrophotography Untracked," copyright 2015, which can be found at https://www.cloudynights.com/articles/cat/articles/dslr-astrophotography-untracked-r3002
Where:
t = maximum exposure time in [seconds]
δ = object declination (degrees)
Δ = pixel size of DSLR chip in [mm]
f = focal length of lens in [mm]
"We used a 50 mm lens @f/2.8 and a 135 mm lens @f/4 with 3 and 1 second exposures, respectively. Except that our Canon 1100D is modified for optimum H-alpha sensitivity, it is an ordinary, off-the-shelf APS-C sized camera body. Although an unmodified DSLR will record less red nebulosity it will still work very well for the method described." Erwin Matys, Karoline Mrazek.

The formula calculates the maximum exposure time without producing star trails. Shorter exposure times can be used. A telescope, depending upon size, may gather more light than a camera telephoto lens. Have fun and experiment with exposure times for both. The exposure limit is somewhere around 5 minutes before star trails begin to appear.

As stated in previous chapters, a sturdy tripod/mount and remote shutter release is required. A programmable intervalometer remote shutter release, like the Vello ShutterBoss II Timer Remote Switch, will automatically take multiple exposures at defined shutter speeds. A camera mirror up function will be of great help in reducing camera shake.

Below are some screen shots of the Maximum Exposure Calculator in the library of www.allocca.com

---

Allocca Biotechnology, LLC
202 East Main Street, Suite 102, Huntington, NY 11743
www.allocca.com

Star Trail Exposure Calculator (Maximum Exposure Time Before Star Trails Without Tracking)

This calulator is based on the work of Erwin Matys and Karoline Mrazek of Project Nightflight.
http://project-nightflight.net/DSLR_astrophotography_untracked.pdf

Lens Focal Length: 50 mm

Object Declination: 60 degrees

Pixel Size: 3.90 microns

Maximum DSLR Exposure Time Before Star Trails Without Tracking:

Assuming an ISO of 6,400 or greater:

Each photo exposure = 4.21 seconds (0.07 minutes)
10 photos will take 42.1 seconds (0.7 minutes) in total
20 photos will take 84.2 seconds (1.4 minutes) in total
30 photos will take 126.3 seconds (2.11 minutes) in total
40 photos will take 168.4 seconds (2.81 minutes) in total
50 photos will take 210.5 seconds (3.51 minutes) in total
60 photos will take 252.6 seconds (4.21 minutes) in total
70 photos will take 294.7 seconds (4.91 minutes) in total
80 photos will take 336.8 seconds (5.61 minutes) in total (Over the 5 Minute Limit)
90 photos will take 378.9 seconds (6.32 minutes) in total (Over the 5 Minute Limit)
100 photos will take 421 seconds (7.02 minutes) in total (Over the 5 Minute Limit)
200 photos will take 842 seconds (14.03 minutes) in total (Over the 5 Minute Limit)

The above times do not indicate the exact exposure time -
Only the Maximum exposure time before star trails appear without tracking.

The resulting star trails will have a length of 2 pixels on the image
2 pixels should not be noticeable, expecially on 4,000 pixel images.

Generally, a maximum exposure is approximately 3-5 minutes. The 500 rule states 500/focal length without taking info consideration pixel size or object declination.

Use an intervalometer shutter release and a sturdy tripod.

---

50 mm FL

Allocca Biotechnology, LLC
202 East Main Street, Suite 102, Huntington, NY 11743
www.allocca.com

Star Trail Exposure Calculator (Maximum Exposure Time Before Star Trails Without Tracking)

This calulator is based on the work of Erwin Matys and Karoline Mrazek of Project Nightflight.
http://project-nightflight.net/DSLR_astrophotography_untracked.pdf

Lens Focal Length: 135 mm

Object Declination: 60 degrees

Pixel Size: 3.90 microns

Maximum DSLR Exposure Time Before Star Trails Without Tracking:

Assuming an ISO of 6,400 or greater:

Each photo exposure = 1.56 seconds (0.03 minutes)
10 photos will take 15.6 seconds (0.26 minutes) in total
20 photos will take 31.2 seconds (0.52 minutes) in total
30 photos will take 46.8 seconds (0.78 minutes) in total
40 photos will take 62.4 seconds (1.04 minutes) in total
50 photos will take 78 seconds (1.3 minutes) in total
60 photos will take 93.6 seconds (1.56 minutes) in total
70 photos will take 109.2 seconds (1.82 minutes) in total
80 photos will take 124.8 seconds (2.08 minutes) in total
90 photos will take 140.4 seconds (2.34 minutes) in total
100 photos will take 156 seconds (2.6 minutes) in total
200 photos will take 312 seconds (5.2 minutes) in total (Over the 5 Minute Limit)

The above times do not indicate the exact exposure time -
Only the Maximum exposure time before star trails appear without tracking.

The resulting star trails will have a length of 2 pixels on the image
2 pixels should not be noticeable, expecially on 4,000 pixel images.

Generally, a maximum exposure is approximately 3-5 minutes. The 500 rule states 500/focal length without taking info consideration pixel size or object declination.

Use an intervalometer shutter release and a sturdy tripod.

135 mm FL

Step by Step Instructions to Record the Photograph

1. Setup camera on tripod, mount, and/or telescope.

2. Set up the camera.

3. Aim the camera / telescope.

4. Remove the lens cap and focus the lens.

5. Click on the ShutterBoss start button.

6. Images should automatically be saved to an SD card.

Note: the mirror function on the Nikon camera works by setting the mirror up on the first press of the shutter release button. The photograph is recorded on the second press of the shutter release button. Therefore, to record 20 images using the ShutterBoss, set the ShutterBoss to release the shutter 40 times.

Step by Step Instructions to Process the Images in Photoshop

File
Scripts
Load files into Stack
Browse for images
Open all files
OK

On the right pane, select all layers

Edit
Auto Align Layers
Select Auto
OK

Edit
Auto-Blend Layers
Select Stack Images
Seamless Tones and Colors should be checked
Content Aware Fill Transparent Areas should be checked
OK

Save the final product accordingly.

Example 1 - Jupiter and Stars With 70 mm Lens

Jupiter, 5-15-17, Nikon D7100, 70 mm, 50 x 3 sec, ISO 6400 Photoshop Processed

Data

Sunset: 8:06 pm

Time of Recording: 903 pm to 9:08 pm

Sky Conditions: Significant amount of light pollution. Only Jupiter and a few stars were visible with the naked eye.

Camera: Nikon D7100 DSLR Camera
  Mirror up function set (MUP)
  Manual Exposure
  Bulb
  Lens Aperture set to 5.0
  ISO 6400

Lens: Nikon AF-S VR Zoom-Nikkor 70-300 mm f/4.5-5.6G
IF-ED
　　　70 mm
　　　Manual Focus

Vello ShutterBoss II Timer Remote Switch for Nikon with
DC2 Connection
　　　3 second time
　　　1 second interval
　　　100 shutter releases (for 50 photos)

Example 2 - Jupiter and Stars With 300 mm Lens

Jupiter, 5-16-17, Nikon D7100, 300mm, 20 x 1 sec, ISO 6400, Photoshop Processed

Data

Sunset: 8:07 pm

Time of Recording: 9:01 pm to 9:03 pm

Sky Conditions: Significant amount of light pollution. Only Jupiter and a few stars were visible with the naked eye.

Camera: Nikon D7100 DSLR Camera
      Mirror up function set (MUP)
      Manual Exposure
      Bulb
      Lens Aperture set to 5.0
      ISO 6400

Lens: Nikon AF-S VR Zoom-Nikkor 70-300 mm f/4.5-5.6G
IF-ED
  300 mm
  Manual Focus

Vello ShutterBoss II Timer Remote Switch for Nikon with
DC2 Connection
  1 second time
  1 second interval
  40 shutter releases (for 20 photos)

Example 3 - Moon With C90 Telescope

Data

Time of Recording: 6/3/17 - 9:20 pm

Sky Conditions: Significant amount of light pollution. Only Jupiter and a few stars were visible with the naked eye.

Camera: Nikon D7100 DSLR Camera
      Mirror up function set (MUP)
      Manual Exposure

1/125
ISO 200

Telescope: Celestron C90 (90 mm x 1,250 mm)

Vello ShutterBoss II Timer Remote Switch for Nikon with DC2 Connection
    Manual (single photo)

Resources

5/8/17

Nikon Cameras

Nikon D3300 DSLR Camera with 18-55mm Lens
24.2MP DX-Format CMOS (crop size) Sensor
15.17 oz (0.95 pounds) body only
6.88 oz lens
No mirror up function
$450

Nikon D7100 DSLR Camera (Body Only)
24.1 MP DX-Format CMOS (crop size) Sensor
23.8 oz. (1.49 pounds) body only
Mirror up function
$700

Nikon D7500 DSLR Camera (Body Only)
20.9MP DX-Format CMOS (crop size) Sensor
Tilting LCD Monitor
22.6 oz. (1.41 pounds) body only
Mirror up function
$1,250

Nikon D750 DSLR Camera (Body Only)
24.3MP FX-Format CMOS (full size) Sensor
Tilting LCD Monitor
26.4 oz. (1.65 pounds) body only
$1,800

DX (crop size) Lenses

Nikon AF-S DX NIKKOR 55-200mm f/4-5.6G ED VR II Lens, 11 oz., $150

Nikon AF-P DX NIKKOR 18-55mm f/3.5-5.6G VR Lens, 6.9 oz., $250

Nikon AF-S DX NIKKOR 55-300mm f/4.5-5.6G ED VR Lens, 18.7 oz., $400

Nikon AF-S DX NIKKOR 18-300mm f/3.5-6.3G ED VR Lens, 19.4 oz., $700

FX (full size) Lenses

Nikon AF-S NIKKOR 24-85mm f/3.5-4.5G ED VR Lens, 16.3 oz., $500

Nikon AF-S NIKKOR 28-300mm f/3.5-5.6G ED VR Lens, 28.2 oz., $950

Nikon AF-S NIKKOR 24-120mm f/4G ED VR Lens, 25 oz., $1100

Nikon Micro-NIKKOR 105mm f/2.8 Lens, 18.1 oz. $730

Sigma 150-600 mm f/5-6.3 DG OS HSM Contemporary Lens for Nikon F (Camera Lens) $1,000

MCT / SCT Telescopes

Celestron C6-A-XLT CG-5 6" f/10 Schmidt-Cassegrain Telescope (OTA)
6" (150 mm) Schmidt-Cassegrain OTA, 8 pounds
1500 mm Focal Length, f/10 Focal Ratio, $600

C90 Maksutov Spotting Scope (OTA)
3.5" (90 mm) Maksutov-Cassegrain OTA, 5 pounds
1250 mm Focal Length, f/13.8 Focal Ratio, $200

Refractor Telescopes

Sky-Watcher ProED 100 mm Doublet APO Refractor (OTA)
100 mm apochromatic Refractor with ED Schott BK-7 and FPL-53 ED glass
900 mm focal length, f/9, 11 pounds, $750

Astro-Tech AT72ED 72 mm f/6 (72 x 432 mm) ED doublet refractor, 5 pounds, $400

Stellarvue SV80 Access - 80 mm Super ED Refractor f/7 (80 x 560 mm) with 2.5" SV Focuser - SV080-ACCESS, 5.5 pounds, $700

Stellarvue SV80 Access - 80 mm Super ED Refractor f/7 (80 x 560 mm) with 2.5" SV Focuser & 2" Diagonal - SV080-ACCESS-DU, 6.4 pounds, $840

Goto Telescopes

Meade ETX90 Observer f/13.8 (Go to) (90 x 1,250 mm) MCT Telescope, 13.3 pounds total, $500

Celestron 6" NexStar Evolution Computerized Telescope - 12090, 35.4 pounds, $1,300

Tripods / Mounts

Davis & Sanford Magnum P343 Aluminum Tripod with Ball Head, 6 pounds, $130

Olivon TR197-16 Tripod with Head (Tall Tripod), 15 pounds, $250

Vixen Porta II Mount Tall, Product Code: 5863TALL, 15 pounds, $400

Manfrotto MT055XPRO3 Aluminum Tripod, 5.5 pounds, $265

Manfrotto 410 Junior Geared Head with 410PL Quick Release Plate, 2.7 pounds, $275

Miscellaneous

Celestron 94009 Lens Shade for C6 and C8 Tubes (Black), Item 94009, $25 (not for C90)

Celestron T-Adapter for Schmidt Cassegrain Telescopes (for C6 SCT), ITEM # 93633-A, $25 (not required for C90)

Tele Vue SLR Prime Focus Camera Adapter for 2" Focusers, ACM-2000, $60

Vello Lens Mount Adapter - T Mount Lens to Nikon F Mount Camera, LANFT, $14

Celestron Vibration Suppression Pads, ITEM # 93503, $50

Vello ShutterBoss II Timer Remote Switch for Nikon with DC2 Connection, $50

Celestron f/6.3 Reducer Corrector for C Series Telescopes, ITEM # 94175, $150

Celestron RACI Illuminated Right Angle Finderscope, ITEM # 93781, $100

Celestron 8-24 mm 1.25" Zoom Eyepiece, Model Number 93230, $80

Lowepro Photo Classic Series BP 300 AW Backpack, $120

Lowepro ProTactic SH 200 AW Camera Shoulder Bag, $130

Mac Sports Collapsible Folding Outdoor Utility Wagon, Blue, $60

Nalgene HDPE Wide Mouth Round Container, 8 oz., 16 oz., $5

A Beginner's Guide to Imaging the Moon

Note: the following chapter "Solar Imaging or Observation with White Light Filtration" also includes a setup for solar and lunar imaging

Telescopes

There are hundreds of choices of telescopes. Refractors provide excellent optics and are usually heavy. Reflectors provide good optics, are large, and require frequent calibration. Catadioptric telescopes are lighter in weight and provide good optics. The larger the aperture, the more light a telescope gathers. The moon is very bright and doesn't require gathering a lot of light. The focal length will determine the size of the object that is projected onto the eyepiece or camera sensor. Some telescopes with long focal lengths will fill the sensor too much and may require a focal reducer. Others won't fill it enough. This article will show various telescopes and focal lengths. One may also choose to use a camera telephoto lens in place of a telescope. There is far more to discuss than this article covers. Photo simulations were derived from the software "Stallarium," which is available free for Mac or PC.

Mounts/Tripods

There are two basic types of mounts, equatorial and alt azimuth (altitude azimuth). Equatorial mounts align with Polaris and follow the path of the stars. Alt azimuth mounts have an up and down and side to side movement. Then, there are computerized mounts that automatically follow the path of the planets and stars. The moon moves somewhat quickly. A computerized mount will follow the moon's path automatically. The moon is about 1/2 degree in diameter and moves 1/2 degree, in about 2 minutes. As a general rule, the moon can be photographed for about 2 seconds before the moon's movement becomes a problem.

Cameras

There are three basic types of cameras. There is the 35 mm DSLR or mirrorless type with either an APS-C (15 mm x 22.5 mm) sensor or full frame (24 mm x 36 mm) sensor. When using a DSLR, the mirror up function should be used to minimize camera shake, a function that is not required on mirrorless cameras. There are also astronomy dedicated video cameras that require a computer to use. Computers are useful when stacking images is required for deep space objects. The moon can be photographed with a single image. Full frame cameras are more expensive than APS-C cameras. When imaging the moon, is the wide field of view obtained from a full frame camera necessary? See the simulations and judge. Also having a camera with an articulating view screen is important for astrophotography.

Exposure

Use a remote switch to prevent camera shaking.
Set the camera for manual focus and exposure.
Use magnified view to fine focus the telescope.
Set the ISO to 200.
Set the shutter speed to 1/125 or 1/250 seconds. This may need to be adjusted faster or slower depending upon the brightness of the moon. A full moon is brighter than a quarter moon.

Focal Multipliers and Focal Reducers

Barlow Lenses multiply the focal length of a telescope. A Barlow lens of 2x indicates using a Barlow lens between the telescope and camera to multiply the focal length by a factor of 2. Often a camera may not have enough "back focus" to use a Barlow lens with a camera. Therefore, they are not recommended in this article.

A focal reducer of f/6.3 indicates using a focal reducer lens between the telescope and camera to reduce the focal length by a factor of 0.63. However, for the telescopes shown herein, the factor is actually f/6.9. Focal reducers also correct and flatten the optics.

Stellarium Simulations

All simulations were done at a field of view of 5.5 degrees using Stellarium software.

Update 4/24/19

Several tests for clarity and color accuracy were performed on the telescopes and camera lens. The C6 Cassegrain telescope was not clear probably due to it not being in proper collimation. Collimation is a process of aligning the mirrors and lenses in a catadioptric telescope. Collimation is not usually required for refractor telescopes. The SV80A refractor telescope was sharp and clear. The color was slightly faded. The 150-600 mm lens was sharp and clear. The color was excellent.

Some will argue that the C6 will provide the best image. However, collimating a telescope is an involved process that could take an hour to perform. Therefore, either the refractor SV80A or SV102A or the 150-600 mm lens is recommended as the best and easiest to use. The 150-600 mm camera lens can also be used for other applications.

Full Frame vs APS-C Sensor Cameras

As seen below, the moon somewhat fills the image plane of the APS-C sensor. The image plane is much larger with the full frame sensor. In this case, the wider field of view of the full frame sensor may be considered a waste of space and money.

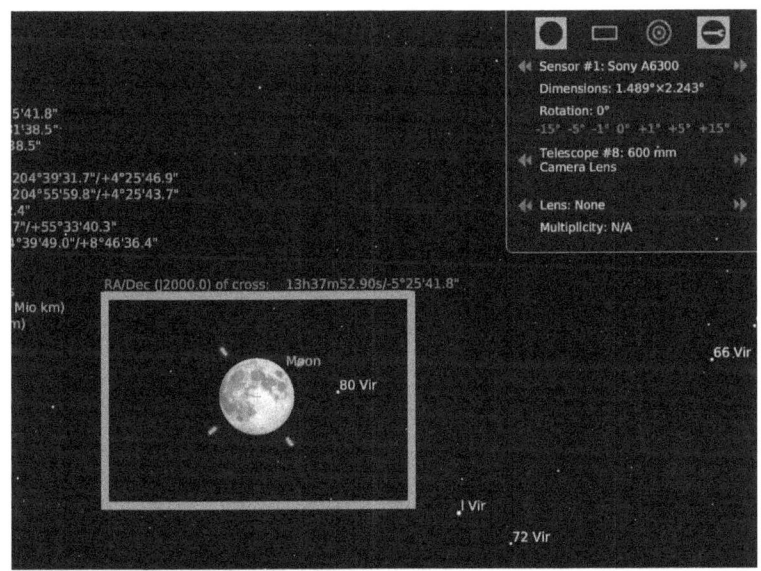

600 mm Camera Lens, APS-C Camera
15 mm x 22.5 mm sensor

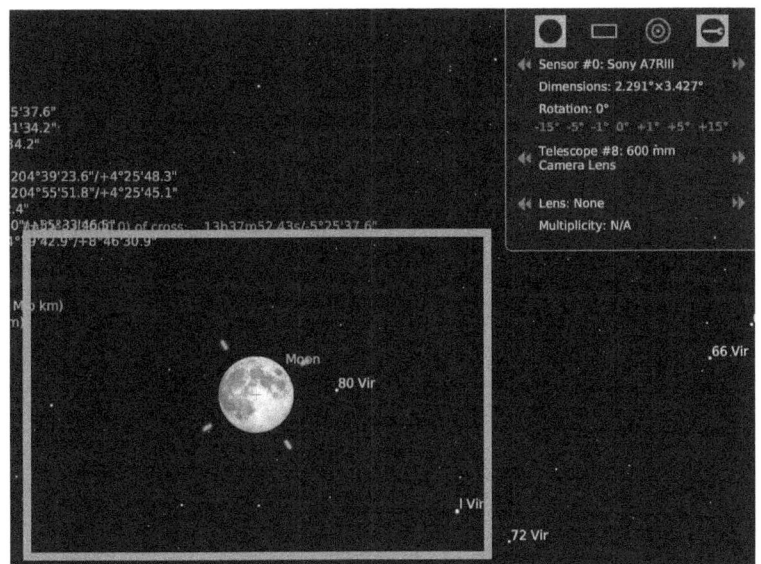

600 mm Camera Lens, Full Frame Camera
24 mm x 36 mm sensor

Telescopes at Various Focal Lengths

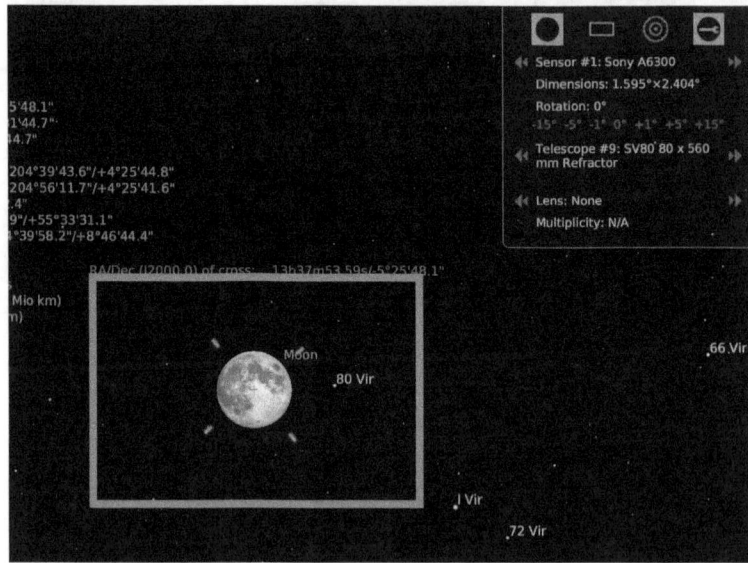

SV80 Refractor, 80 mm x 560 mm, APS-C

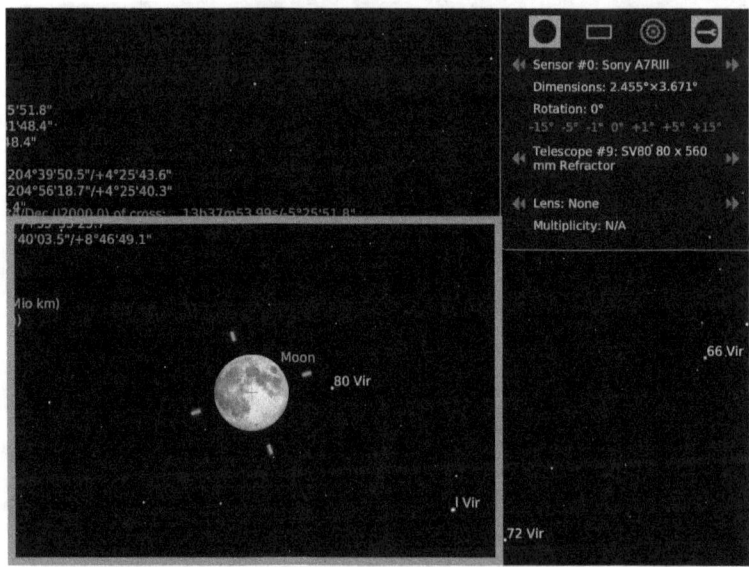

SV80 Refractor, 80 mm x 560 mm, Full Frame

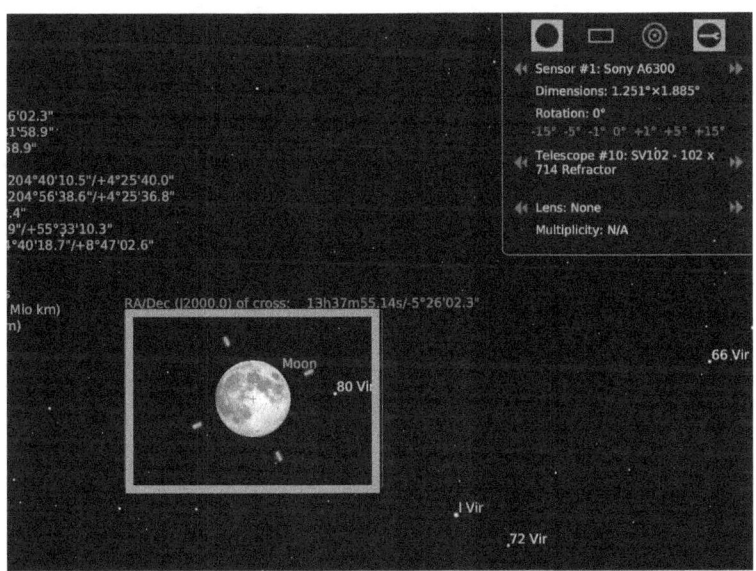

SV102 Refractor, 102 mm x 714 mm, APS-C

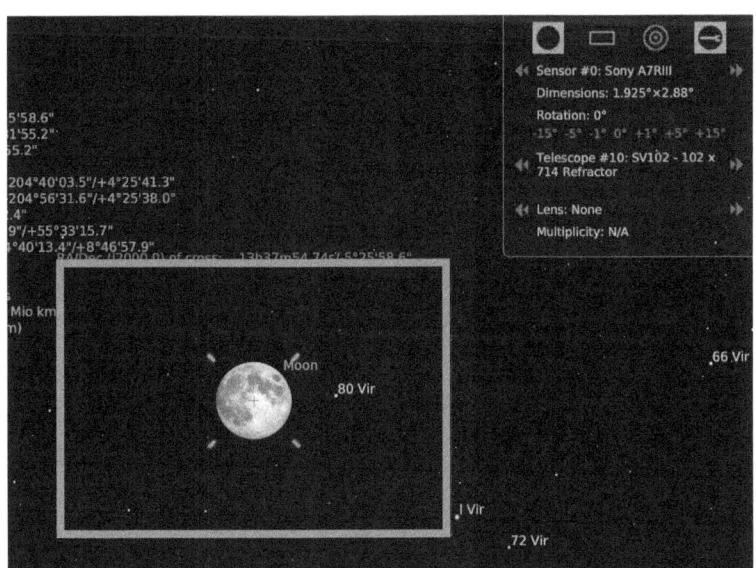

SV102 Refractor, 102 mm x 714 mm, Full Frame

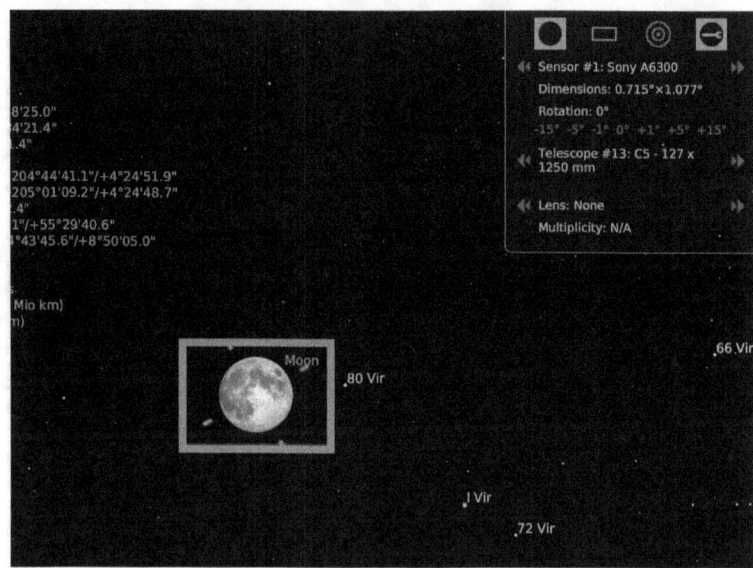

C5 Cassegrain, 127 mm x 1250 mm, APS-C

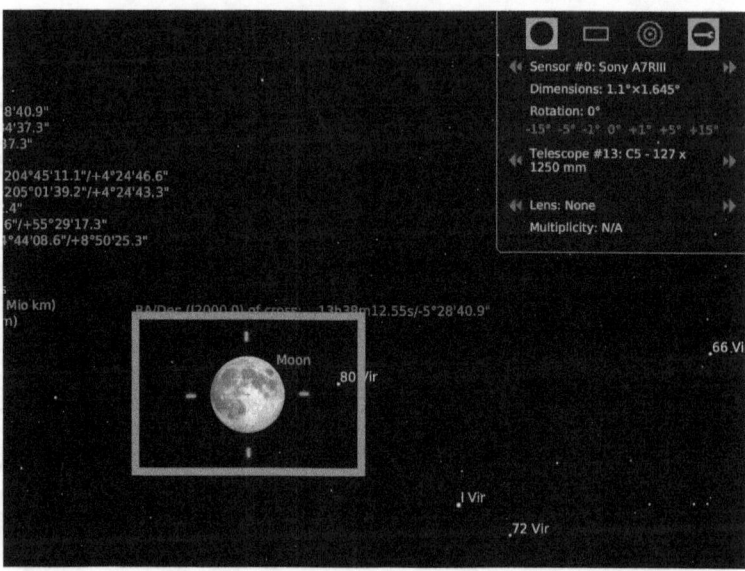

C5 Cassegrain, 127 mm x 1250 mm, Full Frame

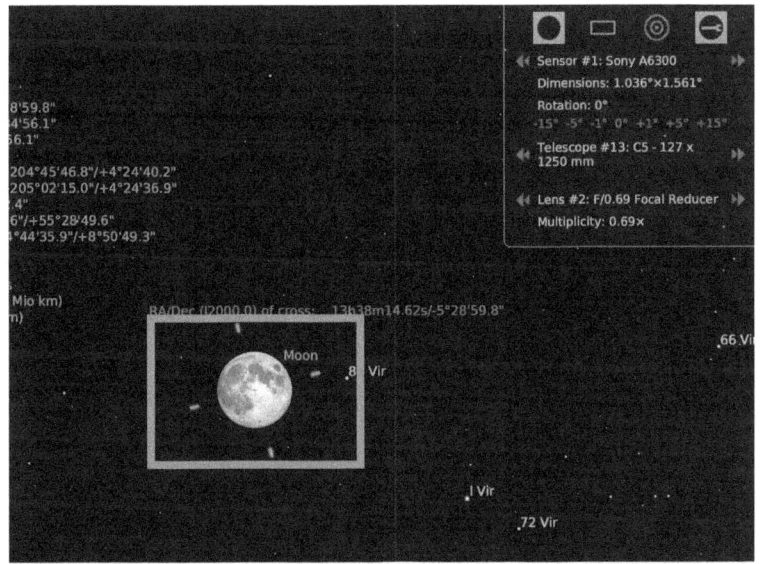

C5, 127 mm x 1250 mm with 0.69 reducer, APS-C

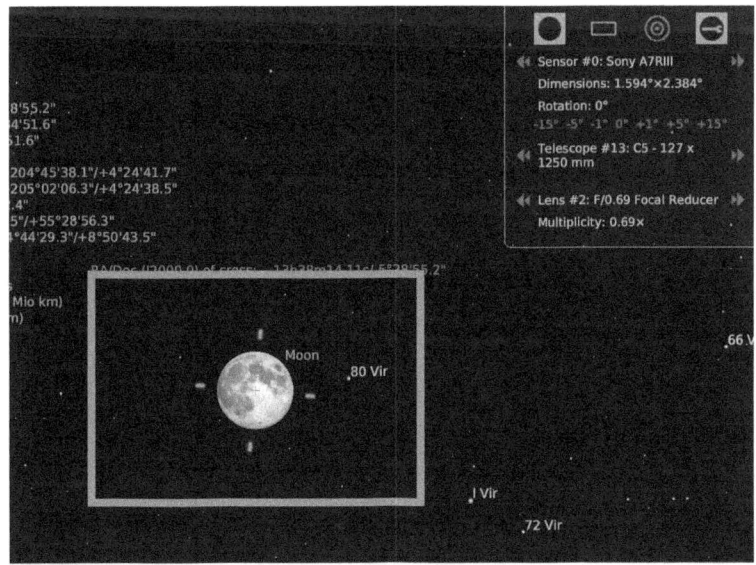

C5, 127 mm x 1250 mm with 0.69 reducer, Full Frame

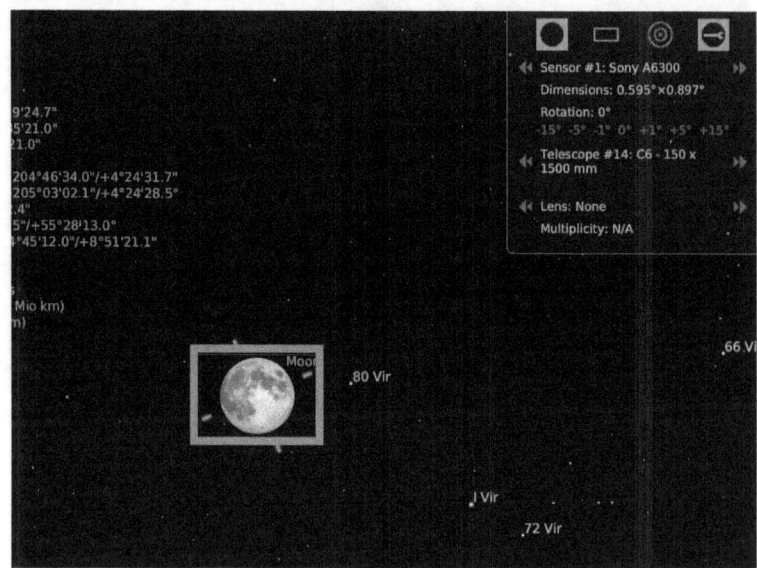

C6, 150 mm x 1500 mm, APS-C

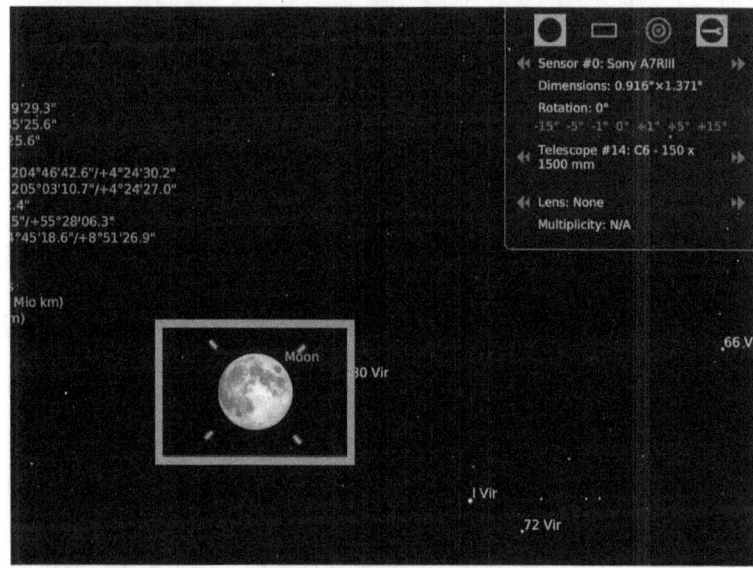

C6 Cassegrain, 150 mm x 1500 mm, Full Frame

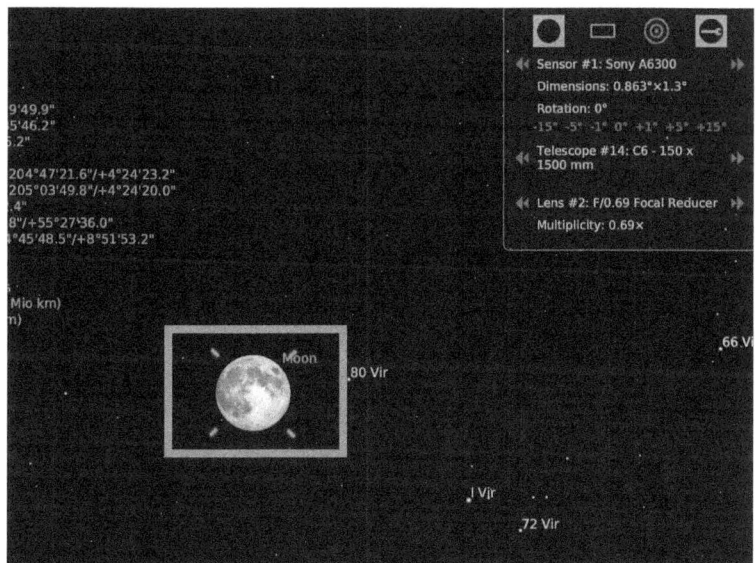

C6, 150 mm x 1500 mm with 0.69 reducer, APS-C

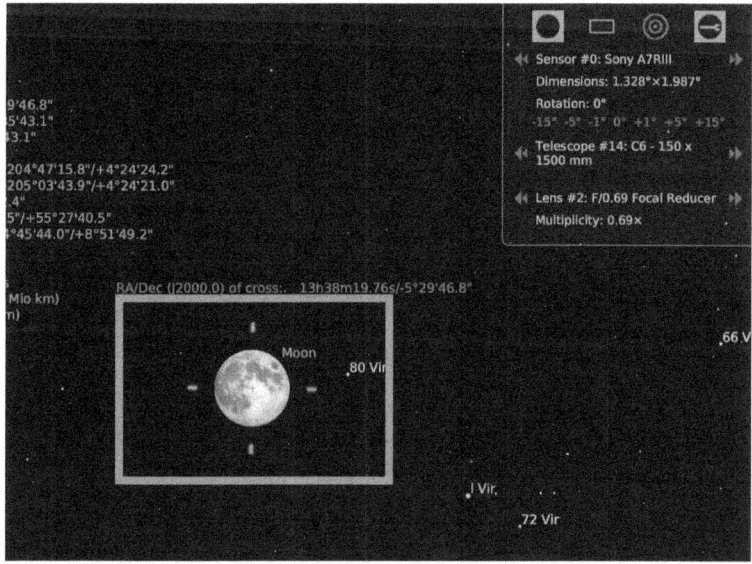

C6, 150 mm x 1500 mm with 0.69 reducer, Full Frame

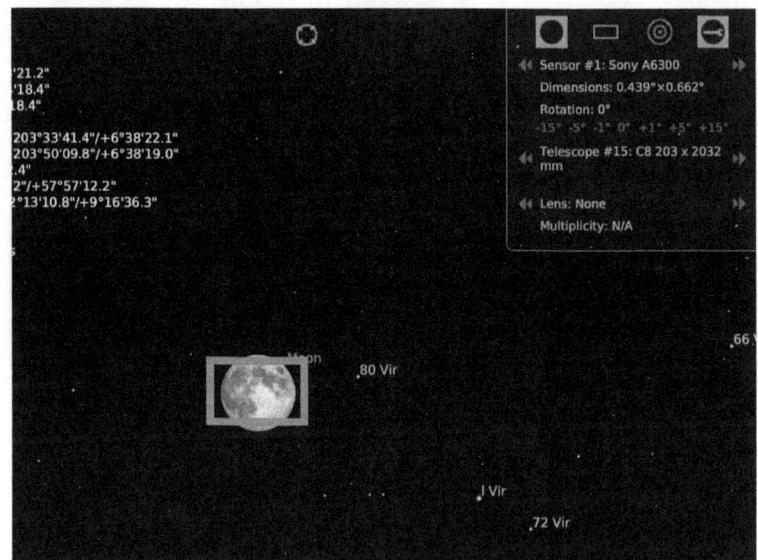

C8 Cassegrain, 203 mm x 2032 APS-C

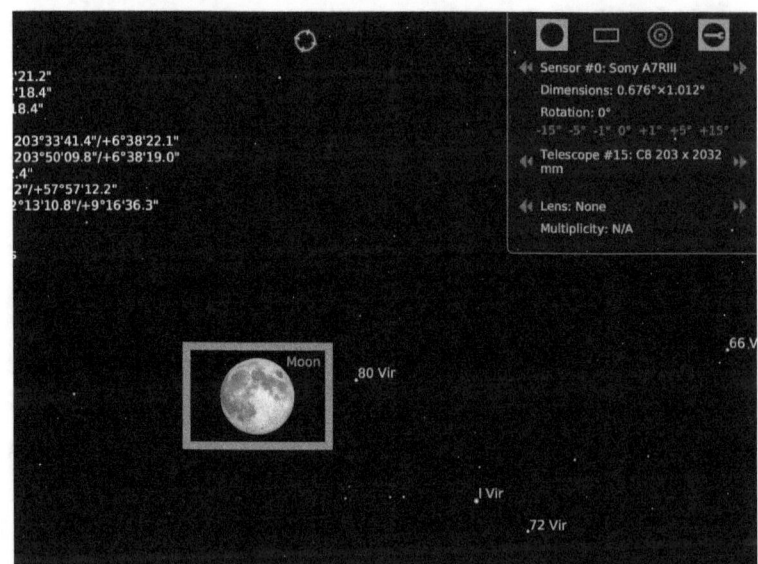

C8 Cassegrain, 203 mm x 2032 Full Frame

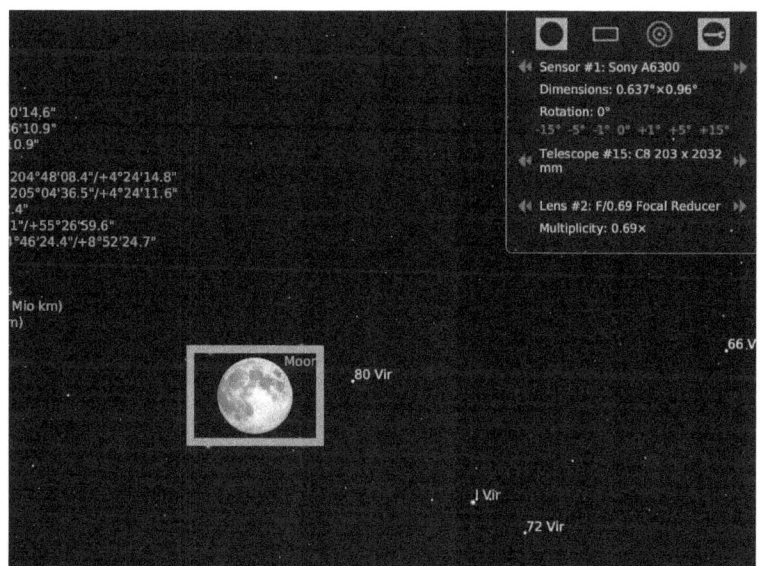

C8, 203 mm x 2032 mm with 0.69 reducer, APS-C

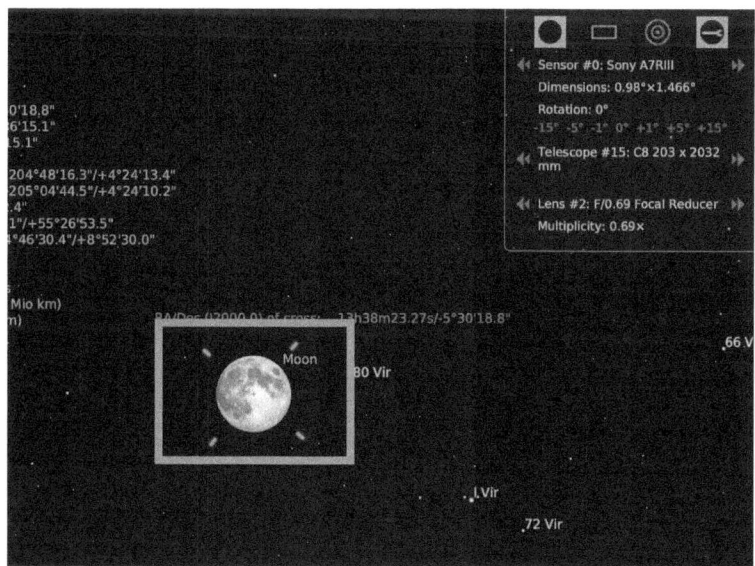

C8, 203 mm x 2032 mm with 0.69 reducer, Full Frame

Comments

The C5 and C6 telescopes without a reducer and the C8 with a reducer fit the frame. However, the moon moves at a somewhat fast rate. It could be out of the frame by the time the camera is focused. A slightly larger field is needed to capture the moon. Too small a telescope or lens will result in significant cropping and enlarging, which results in a loss of resolution.

There are three primary factors in choosing a telescope: function, price, and weight. There are far more choices for telescopes and cameras than are shown here. Refractor telescopes usually provide better optics with a higher weight and cost than Cassegrain telescopes. Other telescopes can work equally as well.

Resources - Celestron C6 Setup

Celestron C6-A-XLT CG-5, 6" f/10 Schmidt-Cassegrain Telescope (OTA)
6" (150 mm) Schmidt-Cassegrain OTA
1500 mm Focal Length, f/10 Focal Ratio
StarBright XLT Optical Coating System
Optical Tube Weight: 10 lbs
List price: $650

Vixen Porta II Mount Tall plus
Adjustable height: 40 to 67 inches
Mount Weight: 15 lb including tripod
Load Capacity: 20 lbs
List Price: $540

Sony Alpha a6300 Mirrorless Digital Camera
24.2 MP DX-Format CMOS Sensor
23.5 x 15.6 mm sensor size
Max Resolution 6000 x 4000
Pixel width = 23.5 / 6000 = 0.0039 = 3.9 microns
Pixel height = 15.6 / 4000 = 0.0039 - 3.9 microns
14.3 oz. (0.89 pounds)
List Price: $750

Fotodiox Lens Mount Adapter, T /T2-Mount Lens to Sony E-Series Camera
Item model number: 11-T-Mount-NEX
List Price: $14

Celestron T-Adapter with SCT 5, 6, 8 with 9.25, 11, 14, Black (93633-A)
Item model number: 93633-A
List Price: $27

Celestron f/6.3 Reducer Corrector for C Series Telescopes
Item model number: 94175
List Price: $150

Vello ShutterBoss II Timer Remote Switch for Sony Multi-Terminal
Item model number: RC-S2II
List Price: $50

Resources - Stellarvue SV80 Access and SV102 Access Setup

Stellarvue SV80 Access - 180 mm Super ED Refractor with 2.5" SV Focuser
Aperture: 80 mm (3.15")
Focal Length: 560 mm
OTA Weight: 6.4 lb
List price: $700

Or

Stellarvue SV102 Access - 102 mm Super ED Refractor with 2.5" SV Focuser
Aperture: 102 mm (4")
Focal Length: 714 mm
OTA Weight: 9.2 lb
List price: $1,100

Vixen Porta II Mount Tall plus
Adjustable height: 40 to 67 inches
Mount Weight: 15 lb including tripod
Load Capacity: 20 lbs
List Price: $540

Sony Alpha a6300 Mirrorless Digital Camera
24.2 MP DX-Format CMOS Sensor
23.5 x 15.6 mm sensor size
Max Resolution 6000 x 4000
Pixel width = 23.5 / 6000 = 0.0039 = 3.9 microns
Pixel height = 15.6 / 4000 = 0.0039 - 3.9 microns
14.3 oz. (0.89 pounds)

List Price: $750

Fotodiox Lens Mount Adapter, T /T2-Mount Lens to Sony E-Series Camera
Item model number: 11-T-Mount-NEX
List Price: $14

Vello ShutterBoss II Timer Remote Switch for Sony Multi-Terminal
Item model number: RC-S2II
List Price: $50

Tele Vue 2" Camera Adapter
Item model number: ACM-2000
List Price: $53

Resources - 150-600 mm Camera and Lens Setup - Sony

Sigma 150-600 mm f/5-6.3 DG OS HSM Contemporary Lens for Canon EF and MC-11 Mount Converter/Lens Adapter for Sony E Kit
16.4 - 4.1 degrees
Filter 95 mm
Minimum Focus Distance 110.2"
Image Stabilized
68.8 oz (4.3 pounds)
$1,000

Sigma MC-11 Mount Converter/Lens Adapter for Sony E Kit
5.15 oz
$250

Sony Alpha a6300 Mirrorless Digital Camera
24.2 MP DX-Format CMOS Sensor
23.5 x 15.6 mm sensor size
Max Resolution 6000 x 4000
Pixel width = 23.5 / 6000 = 0.0039 = 3.9 microns
Pixel height = 15.6 / 4000 = 0.0039 - 3.9 microns
14.3 oz. (0.89 pounds)
List Price: $750

Vello ShutterBoss II Timer Remote Switch for Sony Multi-Terminal
Item model number: RC-S2II
List Price: $50

Manfrotto 502AH Pro Video Head with Flat Base
MVH502AH
Top Attachment: 1/4″ screw, 3/8″ screw
Load Capacity: 15.4 lb
504PLONG Long Quick Release Mounting Plate
Weight: 59.64 oz (3.7 pounds)
$158

Manfrotto 028B Triman Camera Tripod with Geared Center Column
MFR # 028B
Load Capacity: 26.5 lb
Max Height: 89.4"
Min Height: 30.3"
Folded Length: 32.3"
Leg Sections: 3
Weight: 9.2 lb
$353

Note: a less sturdy and expensive tripod and head will likely suffice.

Resources - 150-600 mm Camera and Lens with 1.4x Setup - Nikon

Sigma 150-600 mm f/5-6.3 DG OS HSM Contemporary Lens for Nikon F
16.4 - 4.1 degrees
Filter 95 mm
Minimum Focus Distance 110.2"
Image Stabilized
68.8 oz (4.3 pounds)
$1,000

Nikon D7500 DSLR Camera
Image Sensor: 23.5 x 15.6 mm CMOS sensor
Total Pixels: 21.51 Million
DX-format: 5,568 x 3,712
Pixel width = 23.5 / 5,568 = 0.0042 = 4.2 microns
Pixel height = 15.6 / 3,712 = 0.0042 - 4.2 microns
$800
Weight: 22.6 oz (1.41 pounds)

Sigma TC-1401 1.4x Teleconverter for Nikon F
$350
6.7 oz (0.42 pounds)

Vello Shutterboss Version II Timer Remote Switch for Nikon with DC2 Connection
List Price: $50

Manfrotto 502AH Pro Video Head with Flat Base
MVH502AH
Top Attachment: 1/4″ screw, 3/8″ screw
Load Capacity: 15.4 lb
504PLONG Long Quick Release Mounting Plate
Weight: 59.64 oz (3.7 pounds)
$158

Manfrotto 028B Triman Camera Tripod with Geared Center Column
MFR # 028B
Load Capacity: 26.5 lb
Max Height: 89.4"
Min Height: 30.3"
Folded Length: 32.3"
Leg Sections: 3
Weight: 9.2 lb
$353

Note: a less sturdy and expensive tripod and head will likely suffice.

Note: use the mirror up (MUP) camera mode when imaging the moon.

Photos

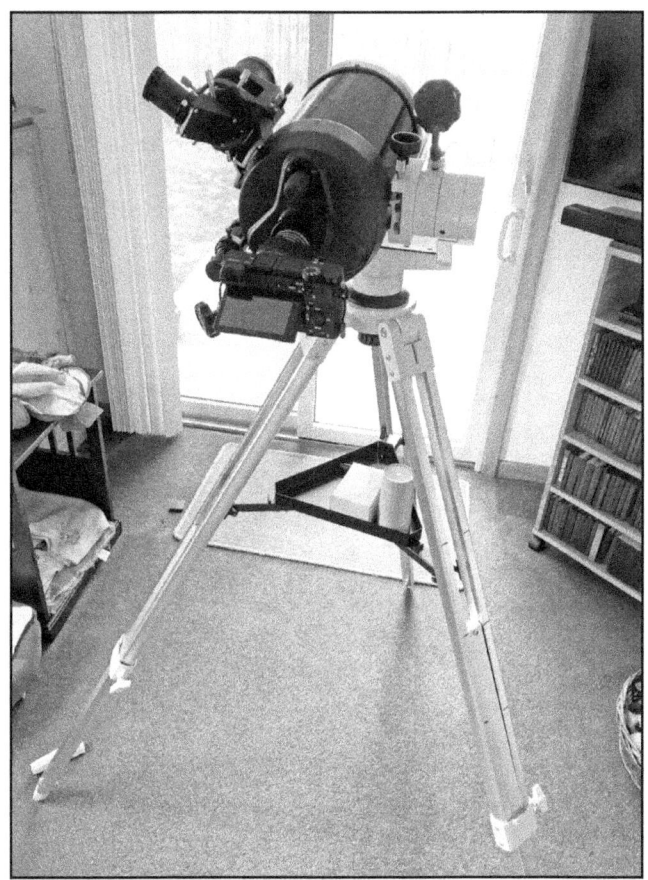

Celestron C6 with Vixen Porta II Tall Mount

Celestron C6 with Camera Adapter, Reducer, and Camera

First Quarter Moon with the Stellarvue SV80 Access
Refractor Telescope (540 mm FL)

Solar Imaging or Observation with White Light Filtration

August 2023

NEVER LOOK DIRECTLY AT THE SUN OR THROUGH A TELESCOPE WITHOUT A SOLAR FILTER

Hydrogen Alpha filtering will provide more detail than white light filtration. However, hydrogen alpha filtering can be very expensive depending upon the system. For this reason, white light filtering was chosen. Neutral density (ND) filters may also be needed to reduce the light and increase contrast depending upon the camera used. The Lunt Herschel wedge has a built in ND filter. There are hundreds of different configurations and at various costs. Below is just one. Alternately, the camera and camera adapter may be replaced by an eyepiece and filters for observation use without imaging.

A Herschel wedge is a right angle prism used to refract most of the light out of the optical path, allowing safe visual observation. It was developed by astronomer John Herschel in the 1830s.

The Telescope Camera Adapter with a T mount specific for Nikon Z cameras allows the camera to connect to the wedge. Other cameras will require a different T mount. The camera adapter chosen allows for magnification by adding an eyepiece inside the adapter. In this setup, a 15 mm eyepiece was used inside the camera adapter. The back focus ability of the telescope will limit the focal length of the eyepiece that can be used. To find the sun, the camera and camera adapter with a 15 mm eyepiece was temporarily replaced with a 40 mm eyepiece onto the Herschel Wedge. After the sun was found and telescope aligned with it, The 40 mm eyepiece was replaced with the camera and adapter containing the 15 mm eyepiece. A Baader Continuum filter may enhance the image, but none were available during the time of this writing.

The Nikon Z5 mirrorless camera used in this configuration has an electronic shutter. Set it to silent mode and use the snap bridge app for remote control so as not to shake the camera using the shutter release button. A shutter speed of at lease 1/2500 and an aperture of f/8 was used. Alternatively, use the interval timer built into the Nikon camera in place of the snap bridge app.

The Manfrotto geared head was used on a tripod. To increase elevation, rotate the head 180 degrees and use it backward.

In figure 3, sun spots can be seen, but not prominences. A hydrogen alpha telescope would be required to see prominences on the sun as shown in the next chapter

The same setup can be used for night time imaging by replacing the Herschel Wedge with a standard 1.25 inch prism or mirror diagonal. The moon will appear the same size as the sun, which is why we can see total eclipses. For a wider field of view, remove the 15 mm eyepiece.

See the chapter "Tracking and Stacking Photos in Photoshop" for higher resolution images.

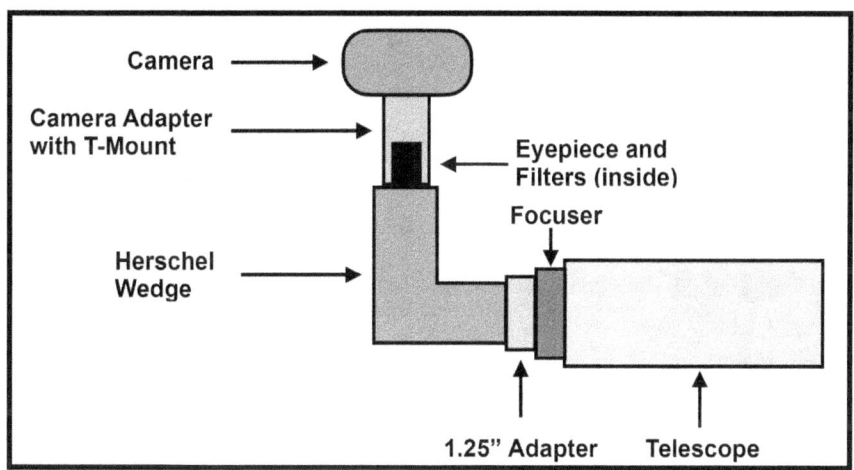

Figure 1 - Telescope Imaging Configuration

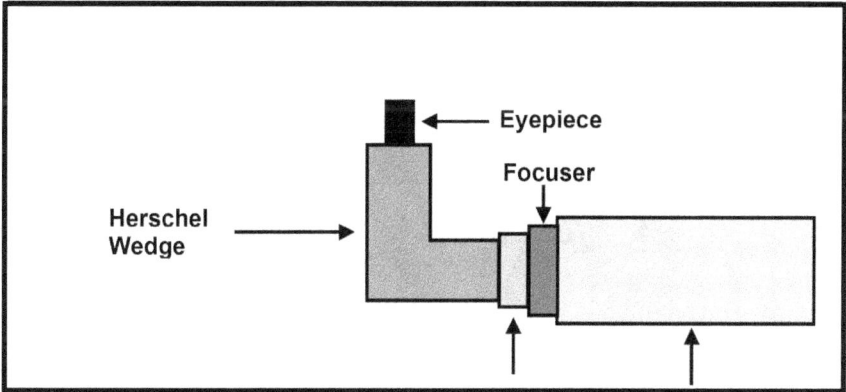

Figure 2 - Telescope Observing Configuration

Figure 3 - Solar Image August 23, 2023, 3:38 pm

Figure 4 - Complete Telescope Setup

This image was photographed through a high quality glass door. Previous tests on trees demonstrated the glass did not affect the quality of the image.

Yes, the sun can be imaged from an air conditioned house as seen from the photo above.

The sun is so large that 190 Earth's side by side would fit across the sun's diameter.

Resources

William Optics Zenithstar 61 mm Doublet Refractor Telescope - Blue Edition, with 1.25" eyepiece adapter
$478

Lunt 1.25" Solar Wedge (Herschel wedge) for refractors
LS1.25HW
$291

SVBONY Telescope Camera Adapter Kit Extendable Prime Focus and Variable Projection Eyepiece Connect 1.25 inches Reflector Telescope Photography
$40

SVBONY Telescope Eyepiece 1.25 inches Telescope Accessory 68 Degree Ultra Wide Angle Astronomy 15 mm
$36

SVBONY Telescope Eyepiece 40mm 1.25 inches Plossl
Telescope Lens Fully Multi Green Coated Metal 40 Degree
Apparent Field 4 Element Telescope Accessory for
Astronomy Telescope
$21

Vello T-Mount Lens to Nikon Z-Mount Camera Lens Adapter
MFR #LA-NZ-T
$13.50

Nikon Z5, 24 mp Mirrorless Camera
$1,295

===
Manfrotto MT055CXPRO4 055 Carbon Fiber 4-Section
Tripod with Horizontal Column
$490

Manfrotto 410 Junior Geared Head with 410PL Quick
Release Plate
$337
===

Or

===
Sky-Watcher AZ-GTi Multi-Purpose Mount & Tripod
Not the best, but inexpensive
===
$475

Or

```
===============================================
```
Sky-Watcher SolarQuest Alt-Azimuth Solar Mount
Not the best, but inexpensive
$530
```
===============================================
```

See the chapter "Tracking and Stacking Photos in Photoshop."

Solar Imaging or Observation with Hydrogen Alpha Filtration

August 2023

NEVER LOOK DIRECTLY AT THE SUN OR THROUGH A TELESCOPE WITHOUT A SOLAR FILTER

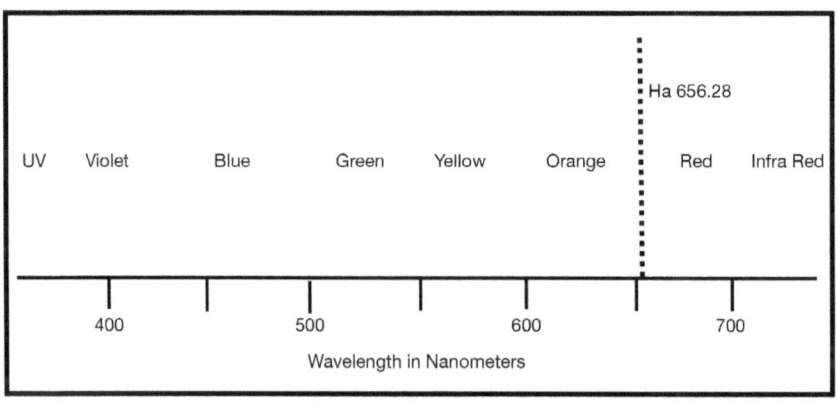

The light from the Sun at the H-alpha frequency (656.28 nanometers) comes from a layer of hydrogen gas that is above the surface of the Sun. This hydrogen layer is called the solar chromosphere. It is only visible through a filter that is within the bandwidths of hydrogen alpha light. The chromosphere contains magnetic energy and it is where the primary solar activity can be observed including filaments, prominences, spicules and active regions. The view is accomplished using the combination of the etalon narrow band filter and hydrogen alpha blocking filter, which blocks all wavelengths of light except hydrogen alpha. There are hundreds of different configurations and at various costs. Below is just one. Alternately, the camera and camera adapter may be replaced by an eyepiece for observation use without imaging.

The Telescope Camera Adapter with a T mount specific for Nikon Z cameras allows the camera to connect to the wedge. Other cameras will require a different T mount. The camera adapter chosen allows for magnification by adding an eyepiece inside the adapter. In this setup, a 15 mm eyepiece was used inside the camera adapter. The back focus ability of the telescope will limit the focal length of the eyepiece that can be used. To find the sun, the camera and camera adapter with a 15 mm eyepiece was temporarily replaced with a 40 mm eyepiece onto the Herschel Wedge. After the sun was found and telescope aligned with it, The 40 mm eyepiece was replaced with the camera and adapter containing the 15 mm eyepiece.

The Nikon Z5 mirrorless camera used in this configuration has an electronic shutter. Set it to silent mode and use the snap bridge app for remote control so as not to shake the camera using the shutter release button. A shutter speed of at lease 1/2500 and an aperture of f/8 was used. Alternatively, use the interval timer built into the Nikon camera in place of the snap bridge app.

The Manfrotto geared head was used on a tripod. To increase elevation, rotate the head 180 degrees and use it backward.

In figure 3, prominences can be seen using a hydrogen alpha telescope.

Also this image was photographed through a high quality glass door. Previous tests on trees demonstrated the glass did not affect the quality of the image.

Yes, the sun can be imaged from an air conditioned house as seen from the photo to right.

Due to some user problems, the Etalon was not adjusted and the imaging session needed to be abandoned. Further sessions were not arranged.

The sun is so large, 190 Earth's side by side would fit across the sun's diameter.

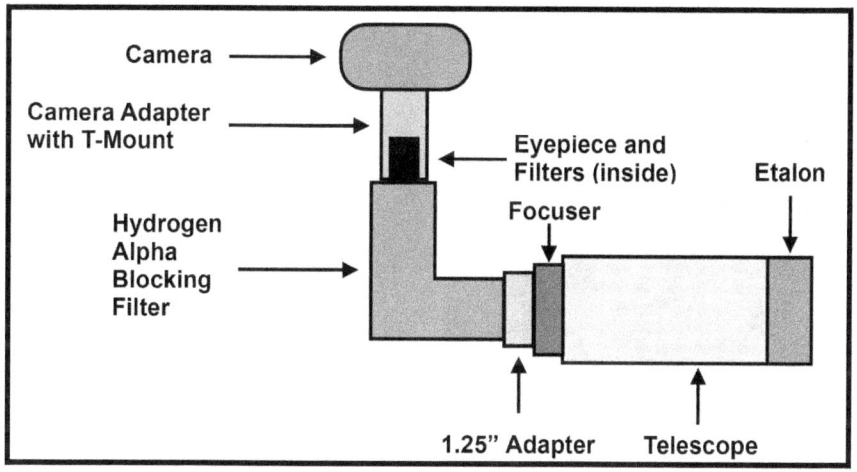

Figure 1 - Telescope Imaging Configuration

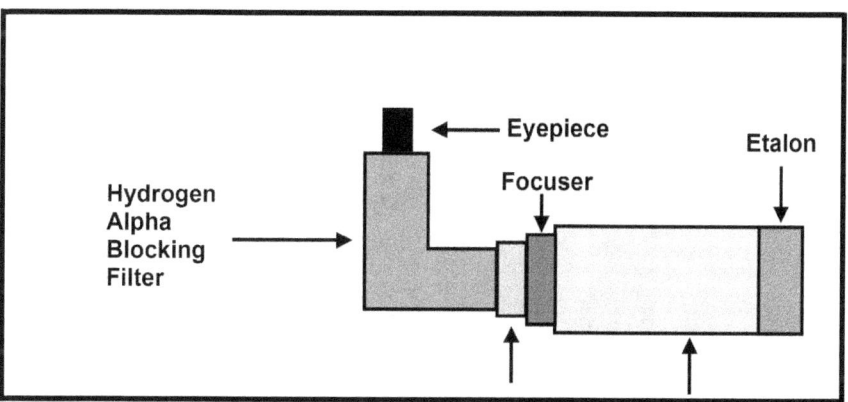

Figure 2 - Telescope Observing Configuration

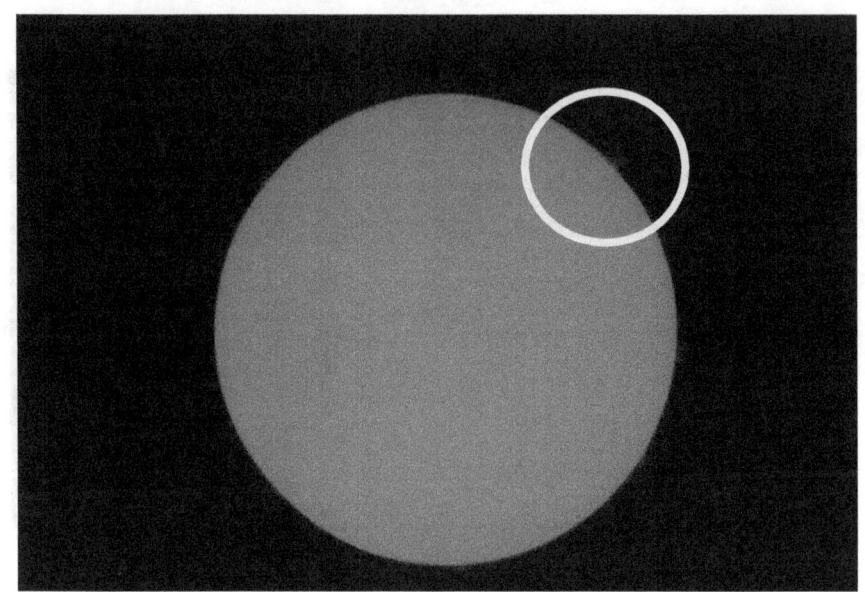

Figure 3 - Solar Image Hydrogen Alpha 8/23/23, 3:38 pm

Figure 4 - Complete Telescope Setup

Resources

Lunt - 40mm Ha Solar Telescope with B600 Blocking filter - Feather Touch Focuser
$1,080

SVBONY Telescope Camera Adapter Kit Extendable Prime Focus and Variable Projection Eyepiece Connect 1.25 inches Reflector Telescope Photography
$40

SVBONY Telescope Eyepiece 1.25 inches Telescope Accessory 68 Degree Ultra Wide Angle Astronomy 15 mm
$36

SVBONY Telescope Eyepiece 40mm 1.25 inches Plossl Telescope Lens Fully Multi Green Coated Metal 40 Degree Apparent Field 4 Element Telescope Accessory for Astronomy Telescope
$21

Vello T-Mount Lens to Nikon Z-Mount Camera Lens Adapter MFR #LA-NZ-T
$13.50

Nikon Z5, 24 mp Mirrorless Camera
$1,295

===
Manfrotto MT055CXPRO4 055 Carbon Fiber 4-Section Tripod with Horizontal Column
$490

Manfrotto 410 Junior Geared Head with 410PL Quick Release Plate
$337

==

Or

==

Sky-Watcher AZ-GTi Multi-Purpose Mount & Tripod
Not the best, but inexpensive

==

$475

Or

==

Sky-Watcher SolarQuest Alt-Azimuth Solar Mount
Not the best, but inexpensive
$530

==

Tracking and Stacking Images

August 2023

The purpose of photo stacking is to improve image resolution and signal to noise ratio. During this process, software will select the sharpest part of the image and add it to the group. Most of the unsharp parts of the image will be left out. There are many different methods of tracking and stacking photos. This is just one.

To image stars, tracking may or may not be necessary. See the chapter "Astrophotography Without Tracking." The 500 rule will provide a quick estimate, which is 500 divided by the focal length in seconds. For example, if the focal length is 250 mm, 500 / 250 = 2 seconds, therefore the image can be tracked for 2 seconds before star trails appear. If a full frame camera is used with a lens, such as 35 mm, the calculation would be 500 / 35 mm = 14.3 seconds.

If eyepiece project is used, it is more complicated. When an eyepiece is added to the camera adapter, it increases the focal length of the telescope.
If a full frame camera is used (every 50 mm = 1x), equation would be:
For example, if telescope focal length is 360 mm and a 15 mm eyepiece is used,

500 / ((360 / 15) x 50)
500 / ((24) x 50)
500 / 1200
= 0.42 seconds to image before star trails appear, which is not much time to acquire many images.

The sun and moon move very quickly. A single image would need to be taken because the sun would be out off the screen in a matter of seconds. If multiple images are taken of the sun and moon tracking would be required. **The higher the magnification, the sturdier the setup will need to be. The slightest micro movement at high magnification even when shooting at 1/2000 of a second can cause a large enough mis-alignment of images whereby the software will not be able to stack them. In this case, a single image may provide better results.**

The best and easiest to use mount in this author's opinion for solar imaging would be the Sky-Watcher SolarQuest Alt-Azimuth Solar Mount. This mount has a sun detector built-in, which detects the sun and aligns the mount automatically. However, during the time of this writing, it was back ordered. The Sky-Watcher AZ-GTi Multi-Purpose Mount & Tripod was used. The AZ-GTi is actually more versatile because it can also be used to image the moon and stars.

In previous older chapters, external intervalometers were recommended. Most newer cameras have built in intervalometers.

Example

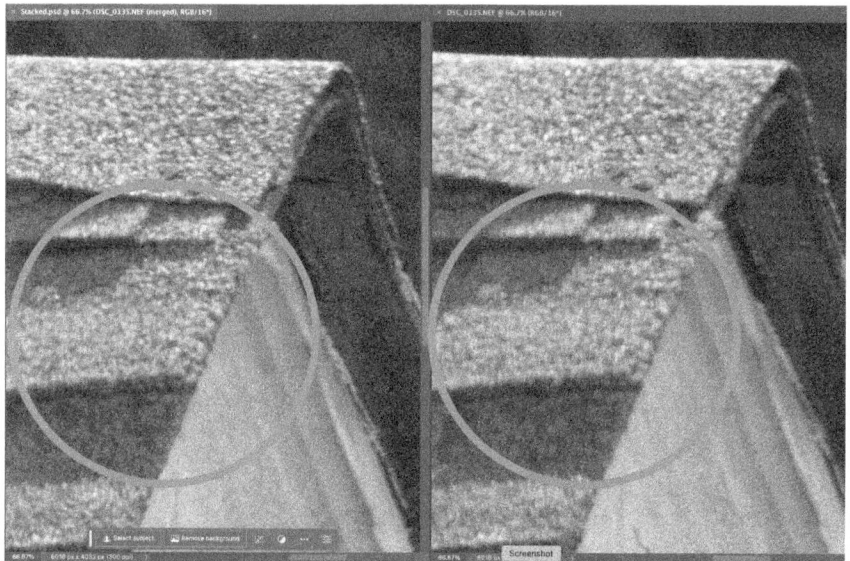

10 images stacked
and enlarged 200%

1 of f10 images
enlarged 200%

The above object was a stationary object as seen through a 61 mm telescope with 15 mm eyepiece projection. This image may be considered a good image. The improvement here is subtle, but can be seen clearly. Some of the grain of the roof tiles on the top of the image was brought out clearly in the stacked image. Therefore stacking can even improve a good image.

Daytime Alignment of the Sky Watcher GTI Alt-Az Mount for Solar Observation and Imaging

Daytime Alignment of Sky Watcher GTI Alt-Az Mount for
Solar Observation and Imaging
9/9/23

Allow Solar Observing

In the Main Menu, go to Settings, then Advanced.

Toggle the "Observe Sun" switch. Two numbers will appear
and a request for the sum.

Enter the sum.

Go back to the Main Menu.

Alignment

Be sure the solar filter is in place on the telescope.

Setup the mount and telescope.

Loosen the clutch on the mount.

Level the telescope

Tighten the clutch on the mount.

Loosen the screw only slightly on the bottom of the extension.

Point the telescope to North.

Tighten the screw on the bottom of the extension. Double check to see if it is level.

Go to the smartphone wifi setting.

Select Synscan wifi.

In the Synscan app, select Connect and AZ Mode or EQ Mode.

Select Alignment on the Synscan app.

Select a one star alignment. The star will not be seen during daylight in the telescope, but click ok anyway.

Click on the star in the right corner of the app.

Select "Solar"

Press the back arrow

Use the up and down and left and right controls on the Synscan app to center the sun. The number in the center indicates the slew rate. Change the slew rate using the right and left arrow keys to the left and right of the controls.

The app should automatically track the sun at this point.

Figure 5 - Sky Watcher GTi Mount and Telescope Setup

Step by Step Instructions to Process Images in Photoshop

Photoshop may not be the best software to stack images, but it is the easiest to use.

File
Scripts
Load files into Stack
Browse for images
Open all files
OK

On the right pane, select all layers

Edit
Auto Align Layers
Select Auto
OK

Edit
Auto-Blend Layers
Select Stack Images
Seamless Tones and Colors should be checked
Content Aware Fill Transparent Areas should be checked
OK

Save the final product accordingly.

Resources

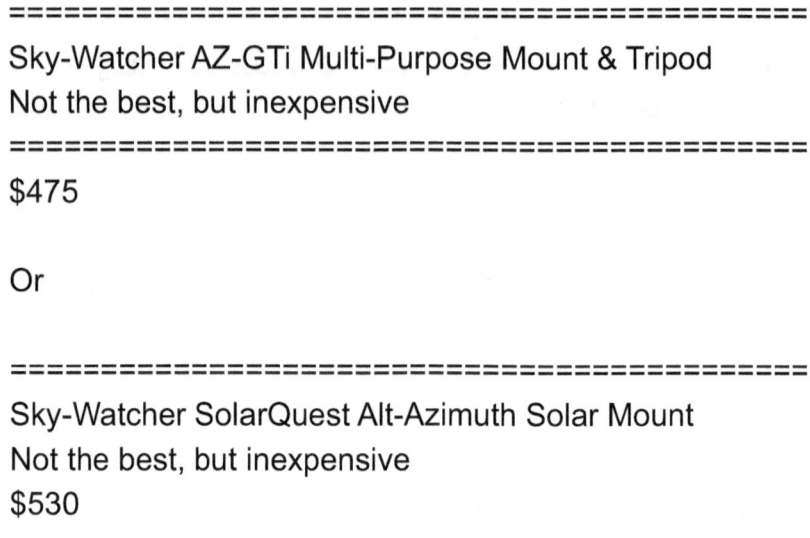

===
Sky-Watcher AZ-GTi Multi-Purpose Mount & Tripod
Not the best, but inexpensive
===
$475

Or

===
Sky-Watcher SolarQuest Alt-Azimuth Solar Mount
Not the best, but inexpensive
$530
===

4Pcs Spirit Bubble Level Mini Square Levels Bullseye Level
Small leveler tool for Picture Frame Hanging RV Tripod
Leveling 40x15x15MM (non-magnetic)
Mount to the telescope with double stick tape.
$11.60

Coghlan's Map Compass
Mount to the telescope with double stick tape.
$6.20

To mount a DSLR or Mirrorless camera and lens in place of a telescope, an L bracket may need to be used. It can be made from the following two parts:

NEEWER 3.15"/80mm Rail Bar Vixen Style Dovetail Plate, Metal Mounting Plate Saddle with 1/4" 3/8" D Ring Screw for

Telescope Mount Adapter Base OTA Equatorial Tripod Sky Astrophotography, QR004
$29.50

PATIKIL L Bracket Tripod Quick Release Plate, Vertical Horizontal Switching Camera Tripod Adapter Mount Parts Replacement Model 2, Black
$16.50

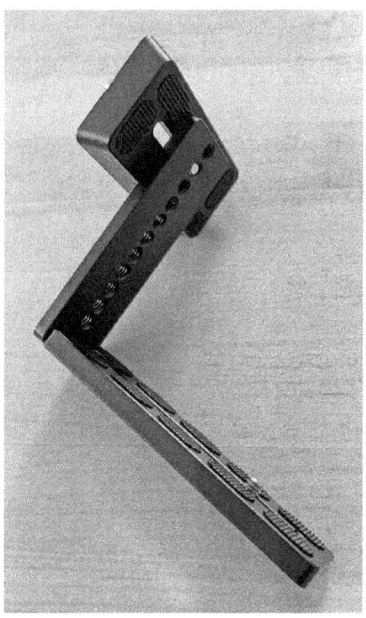

Smart Imaging Telescopes

September 2023

Over the last fews years, smart imaging telescopes with built-in cameras have emerged. Smart imaging telescopes with built-in cameras and processors can be used very easily. The Vaonis Vespera is one example whereby it can be setup on a tripod and leveled. Then, the mobile phone app "Synchronicity" will do the rest.

They can obtain and stack images for long periods of time automatically. At the time of this writing, such telescopes range in price from approximately $500 to $5,000. The more expensive telescopes have larger apertures and larger sensors for higher resolution.

At the time of this writing, the Vaonis Vespera sells for $1,500. Vaonis is releasing a Vespera Pro next year with a higher resolution sensor for $2,500?. Both versions feature a 50 mm quadruplet telescope and weigh less than 10 pounds. Then, there are the usual accessories such as a solar filter, dual band filter, light pollution filter, dew sensor, larger tripod, and backpack. For solar imaging, the solar filter is essential.

Unistellar and ZWO also offer good quality smart imaging telescopes.

www.ingramcontent.com/pod-product-compliance
Lightning Source LLC
Chambersburg PA
CBHW071412170526
45165CB00001B/253